THE STUDY OF HUMAN NATURE

THE STUDY
OF
HUMAN NATURE

Readings selected, edited, and introduced
by

LESLIE STEVENSON

New York　　　　　　Oxford
OXFORD UNIVERSITY PRESS
1981

Copyright © 1981 by Oxford University Press, Inc.

Library of Congress Cataloging in Publication Data
Main entry under title:
The study of human nature.

1. Philosophical anthropology—Addresses, essays
lectures. 2. Man (Theology)—Addresses, essays, lectures. 3. Behaviorism (Psychology)—Addresses, essays,
lectures. I. Stevenson, Leslie Forster.
BD450.S818   128   80-17615
ISBN 0-19-502827-9

Preface

The suggestion for this anthology came from course teachers in the United States who felt a need for a reader to go with my *Seven Theories of Human Nature* (Oxford University Press, 1974). I hope this book will meet that demand, but I also intend it to be readable on its own, whether in college courses or by anyone interested in the subject. The "seven theories" (Plato, Christianity, Marx, Freud, Sartre, Skinner, and Lorenz) are represented here by the most substantial readings, making up about half the total length.

One reviewer of my earlier book (in the *Philosophical Quarterly*, 1975) pointed out, quite rightly, that my original seven are not all "theories" in the same sense. Christianity, in particular, stands out from the rest as a living religion, and insofar as it is *based* on writings (the Bible) they are not the theories of one thinker, but a collection of sacred scriptures of different kinds, being the work of many hands over several centuries. In the cases of Marx and Freud, their work has given rise to continuing political and therapeutic traditions (with their orthodoxies and their heresies), but we have the written theories of a single founder with which to compare contemporary practice. Again, it can be pointed out that by conventional criteria Plato and Sartre are philosophers dealing with metaphysical and ethical themes, whereas Skinner and Lorenz are biological scientists.

However, what made it interesting, I thought, to compare the seven "theories" was that they all could be seen as exemplifying the pattern of theory, diagnosis and prescription—that is, all were concerned not merely to assert some true propositions about human nature, but to pick out certain human problems as of fundamental importance and to offer guidance on how to solve them. There was in each case a promise of liberation from some form of evil or suffering, though it must be admitted that in the cases of Skinner and Lorenz this takes a much less cosmic or metaphysical form than in Christianity, Plato, or even Marx. It may seem inevitable that the more scientific our study of human nature becomes and the more we distinguish empirically testable theories from convictions based on reasoning or intuition alone, the less connection there is between facts and values, the less relevance there is of known truths to how we ought to live. But Freud, Marx, and Sartre offer, in their different ways, analysis of the human condition, which is presented as scientific or philosophical truth, yet which somehow implies a rejection of certain patterns of life and, at the same time, offers the power to change them. Habermas talks of critical or self-reflective studies, whose aim is liberation or emancipation. Perhaps the idea to focus on is not that factual statements may somehow imply value judgments (as so much moral philosophy has denied), but that coming to accept certain truths about ourselves effects a causal change in us. This theme of the liberating power of self-knowledge has, of course, older roots in Plato and Socrates, as well as in Spinoza.

In this anthology, I have tried to place my original seven "theories" in a wider context, while also doing something to distinguish them as belonging to the different categories identified by the rather wordy titles of the four parts of the book. I say a little more about the rationale for this division in the introductory paragraphs for each part. In Part One, I offer a selection of views about human nature from some of the ancient religious traditions, including some brief examples of classical Indian and Chinese thought as well as of biblical writings. But I should emphasize that lack of space, and my own ignorance, have ruled out any attempt to give a representative selection of world religions—that would be a task for a different anthology. In Part Two, with Plato I have also included brief excerpts from other high points of Greek, Roman, and medieval philosophy, which raise themes to be taken up later. Part Three is a large and somewhat mixed bag, loosely unified by what I see as the search for a scientific theory of human nature, from the seventeenth century down to the present day. It has taken that long for substantial results to begin to emerge from the philosophical and methodological perplexities involved in the very idea of a science of human nature. But these problems are far from completely settled, so in Part Four I try to show some of the progress that twentieth-century philosophers have made with them.

There are at least three large philosophical questions that get aired in the course of this anthology. Readers, or course teachers, who want to pursue one of these problems could select readings as follows. The problem of the appropriate *methods* for studying human nature (in particular, whether there must be a difference in principle between physical and social science) is raised in the excerpts from Hume, Mill, Durkheim, Marx, Freud, Watson, Skinner, Lorenz, Chomsky, Davidson, and Habermas. The *mind-body problem* has roots in religion (Hinduism, Buddhism, Christianity), is raised very explicitly in ancient philosophy (all of Part Two), is debated again in the seventeenth century (Descartes, Hobbes, and Spinoza), and is central in twentieth-century philosophy (especially Feigl and Davidson here). The problem of *free will* and its apparent incompatibility with causal explanation is a topic of discussion in Aquinas, Spinoza, Hume, Skinner, Sartre, Hampshire, and Davidson. I would also like to draw attention again to the theme of the liberating power of self-knowledge, which is to be found in the ancient religions and in Plato, Spinoza, Butler, Marx, Freud, Sartre, Hampshire, and Habermas.

The editing has involved more work than might be expected. I have tried to select excerpts that make sense on their own and give the reader who knows nothing of these writers a fair introduction to the main ideas about human nature. In some cases, such as Plato and Freud, the author's exposition is so clear and stylish that the only editorial work needed (or justifiable) was to detach the relevant theories from their original lengthier context. But in other cases, such as those of Marx, Sartre, and Skinner, the length or style of the original presented severe problems for selecting introductory extracts. For them, I have used my editorial scissors very freely, supplied section titles, and sometimes even reordered the excerpts—this is also true for many of the briefer selections.

Clearly, the danger is that, in producing these collages, I may have misrepresented the thought of my victims. Of course, I *hope* I have done them good rather than harm, but given that the original texts are available for anyone who wants to pursue things further, I thought the risk was worth running. For presumably the point of an anthology such as this is to present important ideas as clearly and succinctly as possible, in the hope of communicating them to people who might otherwise never encounter them. A short list of further readings is appended at the end of the book.

My thanks are due to the University of St. Andrews for all that I have been enabled to learn over the past twelve years and for a period of study leave that has facilitated the completion of this work.

Oxford   L.F.S.
April 1980

Contents

PART I BELIEFS ABOUT HUMAN NATURE IN THE
ANCIENT RELIGIOUS TRADITIONS   3

FROM THE OLD TESTAMENT: The creation and the fall
of man; the longing for redemption   5

FROM THE UPANISHADS: Liberation through knowledge
of the true self   11

FROM THE BUDDHIST SCRIPTURES: Introduction; the
doctrine of not-self; the five cardinal virtues;
emancipation and nirvana   17

FROM THE CHINESE CLASSICS: The goodness of
human nature, and the conditions for it to flourish   25

FROM THE NEW TESTAMENT: The redemptive power
of Christ; the resurrection and the life to come   30

PART II REASONED ARGUMENT ABOUT HUMAN
NATURE IN GREEK AND MEDIEVAL
PHILOSOPHY   35

PLATO: The immortality of the soul; the cave, and the way
out of it; justice in society, and the tripartite structure
of the soul; the white and dark horses of the soul   37

LUCRETIUS: The materiality of the soul; the mortality of
the soul; we have nothing to fear from death   60

Contents

ARISTOTLE: Having a soul is being alive; the inseparability
of soul from body; the good for man   66

AQUINAS: The soul is an act of the body, but can exist
separately; the freedom of the will; the purpose of
human life; the resurrection of the body   66

PART III THE SEARCH FOR A SCIENTIFIC THEORY OF HUMAN NATURE   79

RENÉ DESCARTES: Consciousness, and dualism; language
and rationality   81

THOMAS HOBBES: Materialism; conflict between men; the
need for government   86

BARUCH SPINOZA: The mathematical study of human
nature; the identity of mind and body, and the illusion
of freedom; the liberating power of self-knowledge   94

JOSEPH BUTLER: Human nature as a whole integrated by
reflection   102

DAVID HUME: The possibility of an empirical science of
human nature; the compatibility of determinism and
freedom   109

JOHN STUART MILL: That there is, or may be, an inexact
science of human nature; psychology as study of the
laws of mind; the possibility of social science   121

ÉMILE DURKHEIM: The irreducibility of social facts, and
their unavailability to consciousness   131

KARL MARX: Down from heaven to earth; a naturalistic
approach to man; the materialist conception of
history; alienation in capitalist society; communism as
the overcoming of alienation; the need for revolution   136

SIGMUND FREUD: Psychoanalytic theory: id and ego, the
unconscious, and the instincts; repression, neurosis,
and sexuality; psychoanalytic treatment: resistance, the
superego, and transference; the social applications of
psychoanalysis   164

J. B. WATSON: The behaviorist definition of psychology   193

B. F. SKINNER: The application of science to man, and the
threat to ordinary notions of freedom; the external
causes of behavior; the experimental study of
conditioning and reinforcement; the influence of
culture, and the possibility of changing it   199

KONRAD LORENZ: The ethological approach to the
explanation of behavior; innate features in human
social behavior; the evolutionary explanation of
problems in human nature; the problem of human
aggression; the functional limitations of rational
morality   219

NOAM CHOMSKY: The inadequacy of behaviorist accounts
of language; innate factors in language learning; innate
structures in general   237

E. O. WILSON: The genetic determinants in human social
behavior   255

PART IV HOW FAR IS A SCIENTIFIC THEORY OF HUMAN NATURE POSSIBLE? TWENTIETH-CENTURY PHILOSOPHICAL VIEWS   265

JEAN-PAUL SARTRE: Intention and action; nothingness
and freedom; anguish as consciousness of freedom; the
subjectivity of values; bad faith; the inadequacy of
psychoanalytic explanations of bad faith; we are
condemned to be free   267

STUART HAMPSHIRE: Prediction, decision, and freedom
of action; self-knowledge and freedom of thought   297

HERBERT FEIGL: The identity of mental and
neurophysiological events   297

DONALD DAVIDSON: The irreducibility of psychological
to physiological description, and of social to physical
sciences   318

JÜRGEN HABERMAS: The idea of critical social science   325

FOR FURTHER READING   329

THE STUDY OF HUMAN NATURE

Part I

Beliefs about Human Nature
in the
Ancient Religious Traditions

When we go back into the mists of prehistory, looking for the earliest beliefs about human nature, we find a variety of views in the different civilizations and cultures. Certainly, there are elements in common: beliefs in some kind of gods influencing human affairs and in some form of life after death have been very widespread. But there has been enormous variation on these basic themes, and they are not universally present, as we shall see.

The most ancient of all human ideas were handed down from generation to generation by the authority of tradition. As in the few remaining (and fast disappearing) hunter–gatherer societies, there seemed to be no sense of any need to *prove* the world picture of the culture: it was simply what had always been believed and acted upon. Cosmology, religion, and social ethics were rolled up together, typically in a mythlike story which gave a meaning to life in that society, thus forming a primitive kind of ideology. If there was no environmental change, or disruptive contact with other cultures, such aboriginal life could continue with no social or intellectual change over many centuries or even millennia.

Such primitive societies lacked writing and cannot therefore be represented in this anthology. The invention of writing made possible a new stage of intellectual development, because beliefs could then be transmitted across several generations by direct appeal to works that were preserved as sacred or authoritative, and there was also the possibility of adding new elements to the corpus. This is the stage of thought about human nature represented in the following readings. They come from the Middle East (Jewish and Christian), India (Hindu and Buddhist), and China (Confucianist). But there is obviously no attempt here to represent *all* the major religions of the world, only a selection of a few classic writings that have striking things to say about human nature.

Just how or why certain writings acquired authority in their respective religious traditions remains very mysterious. Typically, they contain no attempt to prove, or give evidence for, the doctrines asserted; their concern is much more with conviction and conversion than with epistemology. The beliefs expressed have the status of dogmas and seem to be expected to persuade by their intuitive appeal. Any addition to them, for instance by prophetic inspiration, would have to convince by some such self-authenticating power. One theme in common to all the writings selected here is the hope for some kind of liberation from the suffering, pain, toil, disease, and mortality of ordinary human life. Some traditions are more specific about how such deliverance can be achieved and what form it will take.

From the Old Testament

The Jews were, and are, "the people of the Book." The Old Testament, consisting of the Law (the first five books, attributed to Moses) and the Prophets and other writings, was their guide to life, being accepted as the infallible Word of God. It forms, of course, the first part of the Christian Bible. There is more in it about God than about human nature, because the monotheism of the Hebrews (like that of Islam, another religion that has grown from the same historical root) is awesome in its concentration on the supreme majesty and power of the One Creator and Judge of the universe.

But in the famous creation stories of *Genesis* (Chapters 1 to 3), we *are* presented with beliefs about human nature. Man is said to be "formed of dust from the ground" (suggesting a materialist rather than a dualist view); yet he is made "in the image of God" and is given a special authority over the rest of creation. In the story of the fall, in which the woman for some mysterious reason plays a prominent part, humanity's disobeying God's prohibition leads to the end of primeval innocence, the attainment of knowledge and self-awareness, and condemnation to the pain, toil, and mortality of human existence.

A later stage of Jewish thought is represented here by the classic poetry of *Psalm* 90, attributed to Moses. The emphasis is on the everlasting power of God and, in contrast, on the mortality and sinfulness of man. The longing for God to forgive, and to intervene in history to redeem his people from their plight, is typical of that expressed throughout the Old Testament records of the vicissitudes of the Jews. Christians believe, of course, that this longing is fulfilled in the coming of Jesus. Note that there is no suggestion of human immortality here (and, it seems, little or none in the Old Testament generally); and redemption from sin is more a hope than an offer.

The translation from the Hebrew is that of the Revised Standard Version.

THE CREATION AND THE FALL OF MAN

In the Beginning God created the heavens and the earth. The earth was without form and void, and darkness was upon the face of the deep; and the Spirit of God was moving over the face of the waters. . . .

And God said,"Let the earth put forth vegetation, plants yielding seed, and fruit trees bearing fruit in which is their seed, each according to its kind, upon the earth." And it was so. The earth brought forth vegetation, plants yielding seed according to their own kinds, and trees bearing fruit in which is their seed, each according to its kind. And God saw that it was good. . . .

And God said, "Let the waters bring forth swarms of living creatures, and let birds fly above the earth across the firmament of the heavens."

So God created the great sea monsters and every living creature that moves, with which the waters swarm, according to their kinds, and every winged bird according to its kind. And God saw that it was good. . . .

And God said,"Let the earth bring forth living creatures according to their kinds: cattle and creeping things and beasts of the earth according to their kinds." And it was so.

And God made the beasts of the earth according to their kinds and the cattle according to their kinds, and everything that creeps upon the ground according to its kind. And God saw that it was good.

Then God said,"Let us make man in our image, after our likeness; and let them have dominion over the fish of the sea, and over the birds of the air, and over the cattle, and over all the earth, and over every creeping thing that creeps upon the earth." So God created man in his own image, in the image of God he created him; male and female he created them. And God blessed them, and God said to them, "Be fruitful and multiply, and fill the earth and subdue it; and have dominion over the fish of the sea and over the birds of the air and over every living thing that moves upon the earth."

And God said, "Behold I have given you every plant yielding seed which is upon the face of all the earth, and every tree with seed in its fruit; you shall have them for food.

And to every beast of the earth, and to every bird of the air, and to everything that creeps on the earth, everything that has the breath of life, I have given every green plant for food." And it was so. And God saw everything that he had made, and behold, it was very good. . . .

[T]he Lord God formed man of dust from the ground, and breathed into his nostrils the breath of life; and man became a living being. And the Lord God planted a garden in Eden, in the east; and there he put the man whom he had formed. And out of the ground the Lord God made to grow every tree that is pleasant to the sight and good for food, the tree of life also in the midst of the garden, and the tree of the knowledge of good and evil. . . .

The Lord God took the man and put him in the garden of Eden to till it and keep it. And the Lord God commanded the man, saying, "You may freely eat of every tree of the garden, but of the tree of the knowledge of good and evil you shall not eat, for in the day that you eat of it you shall die." . . .

Out of the ground the Lord God formed every beast of the field and every bird of the air, and brought them to the man to see what he would call them; and whatever the man called every living creature, that was its name. The man gave names to all cattle, and to the birds of the air, and to every beast of the field; but for the man there was not found a helper fit for him. So the Lord God caused a deep sleep to fall upon the man, and while he slept took one of his ribs and closed up its place with flesh; and the rib which the Lord God had taken from the man he made into a woman and brought her to the man. Then the man said,

> "This at last is bone of my bones
>   and flesh of my flesh;
> she shall be called Woman,
>   because she was taken out of Man."

Therefore a man leaves his father and his mother and cleaves to his wife, and they become one flesh. And the man and his wife were both naked, and were not ashamed.

Now the serpent was more subtle than any other wild creature that the Lord God had made. He said to the woman, "Did God say, 'You shall not eat of any tree of the garden'?"

And the woman said to the serpent, "We may eat of the fruit of the trees of the garden; but God said, 'You shall not eat of the fruit of the tree which is in the midst of the garden, neither shall you touch it, lest you die.' " But the serpent said to the woman, "You will not die. For God knows that when you eat of it your eyes will be opened, and you will be like God, knowing good and evil." So when the woman saw that the tree was good for food, and that it was a delight to the eyes, and that the tree was to be desired to make one wise, she took of its fruit and ate; and she also gave some to her husband, and he ate.

Then the eyes of both were opened, and they knew that they were naked; and they sewed fig leaves together and made themselves aprons.

And they heard the sound of the Lord God walking in the garden in the cool of the day, and the man and his wife hid themselves from the presence of the Lord God among the trees of the garden. But the Lord God called to the man, and said to him, "Where are you?" And he said, "I heard the sound of thee in the garden, and I was afraid, because I was naked; and I hid myself." He said, "Who told you that you were naked? Have you eaten of the tree of which I commanded you not to eat?" The man said, "The woman whom thou gavest to be with me, she gave me fruit of the tree, and I ate." The the Lord God said to the woman, "What is this that you have done?" The woman said, "The serpent beguiled me, and I ate." The Lord God said to the serpent,

> "Because you have done this,
>   cursed are you above all cattle,
>   and above all wild animals;
>   upon your belly you shall go,
>   and dust you shall eat
>   all the days of your life.
>   I will put enmity between you and the woman,
>   and between your seed and her seed;
>   he shall bruise your head,
>   and you shall bruise his heel."

To the woman he said,

> "I will greatly multiply your pain in childbearing;
>> in pain you shall bring forth children,
>> yet your desire shall be for your husband,
>> and he shall rule over you."

And to Adam he said,

> "Because you have listened to the voice of your wife,
>> and have eaten of the tree
> of which I commanded you,
>> 'You shall not eat of it,'
> cursed is the ground because of you;
>> in toil you shall eat of it all the days of your life;
> thorns and thistles it shall bring forth to you;
>> and you shall eat the plants of the field.
> In the sweat of your face you shall eat bread
> till you return to the ground,
>> for out of it you were taken;
> you are dust,
>> and to dust you shall return."

THE LONGING FOR REDEMPTION

A Prayer of Moses, the man of God

> Lord, thou hast been our dwelling place
> in all generations.
> Before the mountains were brought forth,
>> or ever thou hadst formed the earth and the world,
> from everlasting to everlasting thou art God.

> Thou turnest man back to the dust, and sayest, "Turn
>> back, O children of men!"
> For a thousand years in thy sight are but as yesterday
>> when it is past,
>> or as a watch in the night.

Thou dost sweep men away; they are like a dream,
   like grass which is renewed in the morning:
in the morning it flourishes and is renewed;
   in the evening it fades and withers.

For we are consumed by thy anger;
   by thy wrath we are overwhelmed.
Thou hast set our iniquities before thee,
   our secret sins in the light of thy countenance.

For all our days pass away under thy wrath,
   our years come to an end like a sigh.
The years of our life are threescore and ten,
   or even by reason of strength fourscore;
yet their span is but toil and trouble;
   they are soon gone, and we fly away.

Who considers the power of thy anger,
   and thy wrath according to the fear of thee?
So teach us to number our days that we may get a heart
   of wisdom.

Return, O Lord! How long?
   Have pity on thy servants!
Satisfy us in the morning with thy steadfast love,
   that we may rejoice and be glad all our days.
Make us glad as many days as thou hast afflicted us,
   and as many years as we have seen evil.
Let thy work be manifest to thy servants,
   and thy glorious power to their children.
Let the favor of the Lord our God be upon us,
   and establish thou the work of our hands upon us,
   yea, the work of our hands establish thou it.

From the Upanishads

The Upanishads are some of the ancient writings of Indian civilization and are included in the sacred scriptures of the Hindu religion. The oldest of them date from about 800–400 B.C.

These extracts from the Katha Upanishad take the form of a dialogue between the boy Nachiketas, who seeks spiritual enlightenment, and a voice symbolizing Death, who delivers increasingly long answers (after some initial reluctance).

In this passage, Death recommends that Nachiketas find the knowledge of true self, the Soul or Spirit, here called the *Atman*. Notice also the brief expression of the typically Indian doctrine of reincarnation, the belief that the soul can be reborn into another body (even trees are imagined as possible recipients), the kind of body depending on the soul's previous performance. The image of the chariot and horses, representing the diverse elements in human nature, recurs in Plato (see Part Two). Human liberation is apparently to be achieved by some form of nonintellectual contemplation or meditation, by which one can reach not only the Atman within oneself, but Brahman, the Supreme Spirit of the universe. (The Hindu religion recognizes a vast multiplicity of gods, but Brahman is supreme.)

The translation from the Sanskrit is by Juan Mascaro, in *The Upanishads* (Penguin Classics, 1965).

LIBERATION THROUGH KNOWLEDGE OF THE TRUE SELF

Nachiketas. When a man dies, this doubt arises: some say "he is" and some say "he is not." Teach me the truth.

Death. Even the gods had this doubt in times of old; for mysterious is the law of life and death. Ask for another boon. Release me from this.

Nachiketas. This doubt indeed arose even to the gods, and you say, O Death, that it is difficult to understand; but no greater teacher than you can explain it, and there is no other boon so great as this.

Death. Take horses and gold and cattle and elephants; choose sons and grandsons that shall live a hundred years. Have vast expanses of land, and live as many years as you desire.

Or choose another gift that you think equal to this, and enjoy it with wealth and long life. Be a ruler of this vast earth. I will grant you all your desires.

Ask for any wishes in the world of mortals, however hard to obtain. To attend on you I will give you fair maidens with chariots and musical instruments. But ask me not, Nachiketas, the secrets of death.

Nachiketas. All these pleasures pass away, O End of all! They weaken the power of life. And indeed how short is all life! Keep thy horses and dancing and singing.

Man cannot be satisfied with wealth. Shall we enjoy wealth with you in sight? Shall we live whilst you are in power? I can only ask for the boon I have asked.

When a mortal here on earth has felt his own immortality, could he wish for a long life of pleasures for the lust of deceitful beauty?

Solve then the doubt as to the great beyond. Grant me the gift that unveils the mystery. This is the only gift Nachiketas can ask.

Death. There is the path of joy, and there is the path of pleasure. Both attract the soul. Who follows the first comes to good; who follows pleasure reaches not the End.

The two paths lie in front of man. Pondering on them, the wise man chooses the path of joy; the fool takes the path of pleasure.

You have pondered, Nachiketas, on pleasures and you have rejected them. You have not accepted that chain of possessions wherewith men bind themselves and beneath which they sink.

There is the path of wisdom and the path of ignorance. They are far apart and lead to different ends. You are, Nachiketas, a follower of the path of wisdom: many pleasures tempt you not.

Abiding in the midst of ignorance, thinking themselves wise and learned, fools go aimlessly hither and thither, like blind led by the blind.

What lies beyond life shines not to those who are childish, or careless, or deluded by wealth. "This is the only world: there is no other," they say; and thus they go from death to death.

Not many hear of him; and of those not many reach him. Wonderful is he who can teach about him; and wise is he who can be taught. Wonderful is he who knows him when taught.

He cannot be taught by one who has not reached him; and he cannot be reached by much thinking. The way to him is through a Teacher who has seen him: He is higher than the highest thoughts, in truth above all thought.

This sacred knowledge is not attained by reasoning; but it can be given by a true Teacher. As your purpose is steady you have found him. May I find another pupil like you!

I know that treasures pass away and that the Eternal is not reached by the transient. I have thus laid the fire of sacrifice of Nachiketas, and by burning in it the transient I have reached the Eternal.

Before your eyes have been spread, Nachiketas, the fulfilment of all desire, the dominion of the world, the eternal reward of ritual, the shore where there is no fear, the greatness of fame and boundless spaces. With strength and wisdom you have renounced them all.

When the wise rests his mind in contemplation on our God beyond time, who invisibly dwells in the mystery of things and in the heart of man, then he rises above pleasures and sorrow.

When a man has heard and has understood and, finding the essence, reaches the Inmost, then he finds joy in the Source of joy. Nachiketas is a house open for thy Atman, thy God.

Nachiketas. Tell me what you see beyond right and wrong, beyond what is done or not done, beyond past and future.

Death. I will tell you the Word that all the *Vedas* glorify, all self-sacrifice expresses, all sacred studies and holy life seek. That Word is OM.

That Word is the everlasting Brahman: that Word is the highest End. When that sacred Word is known, all longings are fulfilled.

It is the supreme means of salvation: it is the help supreme. When that great Word is known, one is great in the heaven of Brahman.

Atman, the Spirit of vision, is never born and never dies. Before him there was nothing, and he is ONE for evermore. Never-born and eternal, beyond times gone or to come, he does not die when the body dies.

If the slayer thinks that he kills, and if the slain thinks that he dies, neither knows the ways of truth. The Eternal in man cannot kill: the Eternal in man cannot die.

Concealed in the heart of all beings is the Atman, the Spirit, the Self; smaller than the smallest atom, greater than the vast spaces. The man who surrenders his human will leaves sorrows behind, and beholds the glory of the Atman by the grace of the Creator.

Resting, he wanders afar; sleeping, he goes everywhere. Who else but my Self can know that God of joy and of sorrows?

When the wise realize the omnipresent Spirit, who rests invisible in the visible and permanent in the impermanent, then they go beyond sorrow.

Not through much learning is the Atman reached, not through the intellect and sacred teaching. It is reached by the chosen of him—because they choose him. To his chosen the Atman reveals his glory.

Not even through deep knowledge can the Atman be reached, unless evil ways are abandoned, and there is rest in the senses, concentration in the mind and peace in one's heart.

Who knows in truth where he is? The majesty of his power carries away priests and warriors, and death itself is carried away. . . .

Know the Atman as Lord of a chariot; and the body as the chariot itself. Know that reason is the charioteer; and the mind indeed is the reins.

The horses, they say, are the senses; and their paths are the objects of sense. When the soul becomes one with the mind and the senses he is called "one who has joys and sorrows."

He who has not right understanding and whose mind is never steady is not the ruler of his life, like a bad driver with wild horses.

But he who has right understanding and whose mind is ever steady is the ruler of his life, like a good driver with well-trained horses.

He who has not right understanding, is careless and never pure, reaches not the End of the journey; but wanders on from death to death.

But he who has understanding, is careful and ever pure, reaches the End of the journey, from which he never returns.

The man whose chariot is driven by reason, who watches and

holds the reins of his mind, reaches the End of the journey, the supreme everlasting Spirit. . . .

I will now speak to you of the mystery of the eternal Brahman; and of what happens to the soul after death.

The soul may go to the womb of a mother and thus obtain a new body. It even may go into trees or plants, according to its previous wisdom and work. . . .

The Tree of Eternity has its roots in heaven above and its branches reach down to earth. It is Brahman, pure Spirit, who in truth is called the Immortal. All the worlds rest on that Spirit and beyond him no one can go:

This in truth is That.

The whole universe comes from him and his life burns through the whole universe. In his power is the majesty of thunder. Those who know him have found immortality.

From fear of him fire burns, and from fear of him the sun shines. From fear of him the clouds and the winds, and death itself, move on their way.

If one sees him in this life before the body passes away, one is free from bondage; but if not, one is born and dies again in new worlds and new creations.

Brahman is seen in a pure soul as in a mirror clear, and also in the Creator's heaven as clear as light; but in the land of shades as remembrance of dreams, and in the world of spirits as reflections in trembling waters.

When the wise man knows that the material senses come not from the Spirit, and that their waking and sleeping belong to their own nature, then he grieves no more.

Beyond the senses is the mind, and beyond mind is reason, its essence. Beyond reason is the Spirit in man, and beyond this is the Spirit of the universe, the evolver of all.

And beyond is Purusha, all-pervading, beyond definitions. When a mortal knows him, he attains liberation and reaches immortality.

His form is not in the field of vision: no one sees him with mortal eyes. He is seen by a pure heart and by a mind and thoughts that are pure. Those who know him attain life immortal.

When the five senses and the mind are still, and reason itself rests in silence, then begins the Path supreme.

This calm steadiness of the senses is called Yoga. Then one should become watchful, because Yoga comes and goes.

Words and thoughts cannot reach him and he cannot be seen by the eye. How can he then be perceived except by him who says "He is"?

In the faith of "He is" his existence must be perceived, and he must be perceived in his essence. When he is perceived as "He is," then shines forth the revelation of his essence.

When all desires that cling to the heart are surrendered, then a mortal becomes immortal, and even in this world he is one with Brahman.

When all the ties that bind the heart are unloosened, then a mortal becomes immortal. This is the sacred teaching.

One hundred and one subtle ways come from the heart. One of them rises to the crown of the head. This is the way that leads to immortality; the others lead to different ends.

Always dwelling within all beings is the Atman, the Purusha, the Self, a little flame in the heart. Let one with steadiness withdraw him from the body even as an inner stem is withdrawn from its sheath. Know this pure immortal light; know in truth this pure immortal light.

And Nachiketas learnt the supreme wisdom taught by the god of after-life, and he learnt the whole teaching of inner-union, of Yoga. Then he reached Brahman, the Spirit Supreme, and became immortal and pure. So in truth will anyone who knows his Atman, his higher Self.

From the Buddhist Scriptures

Buddhism was founded in Northern India about 500 B.C. by the spiritual teacher known as the Buddha; the religion has since spread over the Far East, dividing into different schools. Among the many sacred scriptures, the "Questions of King Milinda" provides one of the easiest introductions to Buddhist thought. In this story, the Buddhist monk Nagasena converts the half-Greek king by his conceptually agile answers to the latter's tricky questions.

The doctrine of reincarnation is defended in parts 2b and 2c of the selected excerpt. Nagasena admits that there is no actual thing that is transferred from one body to another, but thinks that there is some kind of causal connection between them, which is sufficient for us to say that the same person can be reincarnated in a new body, because of that person's previous deeds. Note how Nirvana, the Buddhist ideal of liberation or salvation, is described in negative terms as the cessation of all desires; there is no reference to any God or Supreme Spirit.

These extracts are from the translation by Edward Conze, in *Buddhist Scriptures* (Penguin Classics, 1959).

1. INTRODUCTION

The Venerable Nagasena stayed at the Sankheyya hermitage together with 80,000 monks. King Milinda, accompanied by a retinue of 500 Greeks, went up to where he was, gave him a friendly and courteous greeting, and sat on one side. Nagasena returned his greetings, and his courtesy pleased the king's heart.

2. THE DOCTRINE OF NOT-SELF

2a. The Chariot

And King Milinda asked him: "How is your Reverence known, and what is your name, Sir?" "As Nagasena I am known, O great king, and as Nagasena do my fellow religious habitually address me. But although parents give such names as Nagasena, or Surasena, or Virasena, or Sihasena, nevertheless this word 'Nagasena' is just a denomination, a designation, a conceptual term, a current appellation, a mere name. For no real person can here be apprehended." But King Milinda explained: "Now listen, you 500 Greeks and 80,000 monks, this Nagasena tells me that he is not a real person! How can I be expected to agree with that!" And to Nagasena he said: "If, most reverend Nagasena, no person can be apprehended in reality, who then, I ask you, gives you what you require by way of robes, food, lodging, and medicines? Who is it that consumes them? Who is it that guards morality, practices meditation, and realizes the [four] Paths and their Fruits, and thereafter Nirvana? Who is it that kills living beings, takes what is not given, commits sexual misconduct, tells lies, drinks intoxicants? Who is it that commits the five Deadly Sins? For, if there were no person, there could be no merit and no demerit; no doer of meritorious or demeritorious deeds, and no agent behind them; no fruit of good and evil deeds, and no reward or punishment for them. If someone should kill you, O Venerable Nagasena, he would not commit any murder. And you yourself, Venerable Nagasena, would not be a real teacher, or instructor, or ordained monk! You just told me that your fellow religious habitually address you as 'Nagasena.' What then is this 'Nagasena'? Are perhaps the hairs of the head 'Nagasena'?—"No, great king!" "Or perhaps the hairs of the body?"—"No, great king!" "Or perhaps the nails, teeth, skin, muscles, sinews, bones, marrow, kidneys, heart, liver, serous membranes, spleen, lungs, intestines, mesentery, stomach, excrement, the bile, phelgm, pus, blood, grease, fat, tears, sweat, spittle, snot, fluid of the joints, urine, or the brain in the skull—are they this 'Nagasena'?"—"No, great king!"—"Or is form this 'Nagasena,' or feeling, or perceptions, or impulses, or consciousness?"—"No, great king!"—"Then is it the combination of form, feelings, perceptions, impulses, and consciousness?"—"No, great king!"—"Then is it outside the combination of form, feelings, perceptions, impulses, and consciousness?"—"No, great king!"— "Then, ask as I may, I can discover no Nagasena at all. Just a mere sound is this 'Nagasena,' but who is the real Nagasena? Your Reverence has told a lie, has spoken a falsehood! There really is no Nagasena!"

Thereupon the Venerable Nagasena said to King Milinda: "As a king you have been brought up in great refinement and you avoid roughness of any kind. If you would walk at midday on this hot, burning, and sandy ground, then your feet would have to tread on the rough and gritty gravel and pebbles, and they would hurt you, your body would get tired, your mind impaired, and your awareness of your body would be associated with pain. How then did you come—on foot, or on a mount?"

"I did not come, Sir, on foot, but on a chariot."—"If you have come on a chariot, then please explain to me what a chariot is. Is the pole the chariot?"—"No, reverend Sir!"—"Is then the axle the chariot?"—"No, reverend Sir!"—"Is it then the wheels, or the framework, or the flag-staff, or the yoke, or the reins, or the goad-stick?"—"No, reverend Sir!"—"Then is it the combination of pole, axle, wheels, framework, flag-staff, yoke, reins, and goad which is the 'chariot'?"—"No, reverend Sir!"—"Then is this 'chariot' outside the combination of pole, axle, wheels, framework, flag-staff, yoke, reins, and goad?"—"No, reverend Sir!"—"Then, ask as I may, I can discover no chariot at all. Just a mere sound is this 'chariot'. But what is the real chariot? Your Majesty has told a lie, has spoken a falsehood! There really is no chariot! Your Majesty is the greatest king in the whole of India. Of whom then are you afraid, that you do not speak the truth?" And he exclaimed: "Now listen, you 500 Greeks and 80,000 monks, this king Milinda tells me that he has come on a chariot. But when asked to explain to me what a chariot is, he cannot establish its existence. How can one possible approve of that?"

The five hundred Greeks thereupon applauded the Venerable Nagasena and said to king Milinda: "Now let your Majesty get out of that if you can!"

But king Milinda said to Nagasena: "I have not, Nagasena, spoken a falsehood. For it is in dependence on the pole, the axle, the wheels, the framework, the flag-staff, etc., that there takes place this denomination 'chariot,' this designation, this conceptual term, a current appeliation and a mere name."—"Your Majesty has spoken well about the chariot. It is just so with me. In dependence on the thirty-two parts of the body and the five Skandhas there takes place this denomination 'Nagasena,' this designation, this conceptual term, a current appellation and a mere name. In ultimate reality, however, this person cannot be apprehended. And this has been said by our Sister Vajira when she was face to face with the Lord:

> 'Where all constituent parts are present,
> The word "a chariot" is applied.
> So likewise where the skandhas are,
> The term a "being" commonly is used.'

"It is wonderful, Nagasena, it is astonishing, Nagasena! Most brilliantly have these questions been answered! Were the Buddha himself here, he would approve what you have said. Well spoken, Nagasena, well spoken!"

2b. *Personal Identity and Rebirth*

The king asked: "When someone is reborn, Venerable Nagasena, is he the same as the one who just died, or is he another?"—The Elder replied: "He is neither the same nor another."—"Give me an illustration!"—"What do you think, great king: when you were a tiny infant, newly born and quite soft, were you then the same as the one who is now grown up?"—"No, that infant was one, I, now grown up, am another."—"If that is so, then, great king, you have had no mother, no father, no teaching, and no schooling! Do we then take it that there is one mother for the embryo in the first stage, another for the second stage, another for the third, another for the fourth, another for the baby, another for the grown-up man? Is the schoolboy one person, and the one who has finished school another? Does one commit a crime, but the hands and feet of another are cut off?"—"Certainly not! But what would you say, Reverend Sir, to all that?"—The Elder replied: "I was neither the tiny infant, newly born and quite soft, nor am I now the grown-up man; but all these are comprised in one unit depending on this very body."—"Give me a simile!"—"If a man were to light a lamp, could it give light throughout the whole night?"—"Yes, it could."—"Is now the flame which burns in the first watch of the night the same as the one which burns in the second?"—"It is not the same."—"Or is the flame which burns in the second watch the same as the one which burns in the last one?"—"It is not the same."—"Do we then take it that there is one lamp in the first watch of the night, another in the second, and another again in the third?"—"No, it is because of just that one lamp that the light shines throughout the night."—"Even so must we understand the collocation of a series of successive dharmas. At rebirth one dharma arises, while another stops; but the two processes take place almost simultaneously (i.e. they are continuous). Therefore the first act of consciousness in the new existence is neither the same as the last act of consciousness in the previous existence, nor is it another."—"Give me another simile!"—"Milk, once the milking is done, turns after some time into curds; from curds it turns into fresh butter, and from fresh butter into ghee. Would it now be correct to say that the milk is the same thing as the curds, or the fresh butter, or the ghee?"—"No, it would not. But they have been produced because of it."—"Just so must be understood the collocation of a series of successive dharmas."

2c. *Personal Identity and Karma*

The king asked: "Is there, Nagasena, any being which passes on from this body to another body?"—"No, your majesty!"—"If there were no passing on from this body to another, would not one then in one's next life be freed from the evil deeds committed in the past?"—"Yes, that would be so if one were not linked once again with a new organism. But since, your majesty, one is linked again with a new organism, therefore one is not freed from one's evil deeds."—"Give me a simile!"—"If a man should steal another man's mangoes, would he deserve a thrashing for that?"—"Yes, of course!"—"But he would not have stolen the very same mangoes as the other one had planted. Why then should he deserve a thrashing?"—"For the reason that the stolen mangoes had grown because of those that were planted."—"Just so, your majesty, it is because of the deeds one does, whether pure or impure, by means of this psycho-physical organism, that one is once again linked with another psycho-physical organism, and is not freed from one's evil deeds."—"Very good, Nagasena!"

3. THE FIVE CARDINAL VIRTUES

The king said: "Is it through wise attention that people become exempt from further rebirth?"—"Yes, that is due to wise attention, and also to wisdom, and the other wholesome dharmas."—"But is not wise attention the same as wisdom?"—"No, your majesty! Attention is one thing, and wisdom another. Sheep and goats, oxen and buffaloes, camels and asses have attention, but wisdom they have not."—"Well put, Nagasena!"

The king said: "What is the mark of attention, and what is the mark of wisdom?"—"Consideration is the mark of attention, cutting off that of wisdom."—"How is that? Give me a simile!"—"You know barley-reapers, I suppose?"—"Yes, I do."—"How then do they reap the barley?"—"With the left hand they seize a bunch of barley, in the right hand they hold a sickle, and they cut the barley off with that sickle."—"Just so, your majesty, the Yogin seizes his mental processes with his attention, and by his wisdom he cuts off the defilements."—"Well put, Venerable Nagasena!"

The king said: "When you just spoke of 'the other wholesome dharmas,' which ones did you mean?"—"I meant morality, faith, vigour, mindfulness, and concentration."—"And what is the mark of morality?"—"Morality has the mark of providing a basis for all wholesome dharmas, whatever they may be. When based on morality, all the wholesome dharmas will not dwindle away."—"Give me an illustration!"—"As all plants and animals which increase, grow, and prosper, do so with the earth as their support, with the earth as their basis, just so the Yogin, with morality as his support, with morality as his basis, develops the five cardinal virtues, i.e. the cardinal virtues of faith, vigour, mindfulness, concentration, and wisdom.". . .

4. EMANCIPATION AND NIRVANA

4a. Problems of Nirvana

The king asked: "Is cessation Nirvana?"—"Yes, your majesty!"—
"How is that, Nagasena?"—"All the foolish common people take
delight in the senses and their objects, are impressed by them, are
attached to them. In that way they are carried away by the flood,
and are not set free from birth, old age, and death, from grief,
lamentation, pain, sadness, and despair—they are, I say, not set free
from suffering. But the well-informed holy disciples do not take
delight in the senses and their objects, are not impressed by them,
are not attached to them, and in consequence their craving ceases;
the cessation of craving leads successively to that of grasping, of
becoming, of birth, of old age and death, of grief, lamentation,
pain, sadness, and despair—that is to say to the cessation of all this
mass of ill. It is thus that cessation is Nirvana.". . .

4b. The Nature of Nirvana

King Milinda said: "I will grant you, Nagasena, that Nirvana is
absolute Ease, and that nevertheless one cannot point to its form or
shape, its duration or size, either by simile or explanation, by reason or by argument. But is there perhaps some quality of Nirvana
which it shares with other things, and which lends itself to a meta-
phorical explanation?"—"Its form, O king, cannot be elucidated by
similes, but its qualities can." "How good to hear that, Nagasena!
Speak then, quickly, so that I may have an explanation of even one
of the aspects of Nirvana! Appease the fever of my heart! Allay it
with the cool sweet breezes of your words!"
    "Nirvana shares one quality with the lotus, two with water,
three with medicine, ten with space, three with the wishing jewel,
and five with a mountain peak. As the lotus is unstained by water,
so is Nirvana unstained by the defilements.—As cool water allays
feverish heat, so also Nirvana is cool and allays the fever of all the

passions. Moreover, as water removes the thirst of men and beasts who are exhausted, parched, thirsty, and overpowered by heat, so also Nirvana removes the craving for sensuous enjoyments, the craving for further becoming, the craving for the cessation of becoming.—As medicine protects from the torments of poison, so Nirvana from the torments of the poisonous passions. Moreover, as medicine puts an end to sickness, so Nirvana to all sufferings. Finally, Nirvana and medicine both give security.—And these are the ten qualities which Nirvana shares with space. Neither is born, grows old, dies, passes away, or is reborn; both are unconquerable, cannot be stolen, are unsupported, are roads respectively for birds and Arhats to journey on, are unobstructed and infinite.—Like the wishing jewel, Nirvana grants all one can desire, brings joy, and sheds light.—As a mountain peak is lofty and exalted, so is Nirvana. As a mountain peak is unshakeable, so is Nirvana. As a mountain peak is inaccessible, so is Nirvana inaccessible to all the passions. As no seeds can grow on a mountain peak, so the seeds of all the passions cannot grow in Nirvana. And finally, as a mountain peak is free from all desire to please or displease, so is Nirvana."— "Well said, Nagasena! So it is, and as such I accept it."

From the Chinese Classics

The thought of ancient China seems to have been less obviously religious, and more man-centered, than that of Palestine or India. The *Analects* of Confucius contain much about ethics and politics, and only a litttle about "Heaven." There is more about human nature in the sayings attributed to Mencius (Meng K'o), a teacher in the Confucian tradition who lived in the fourth century B.C.

The selected sayings clearly express an optimistic view of the natural goodness of man, and a concern with the conditions that will allow this goodness to flourish (in "human-heartedness"). The recommendation to "nourish the vast-flowing vital energy in man" may be somewhat vague, but the absence of all reference to any God or Soul or life after death, or even Buddhist renunciation, suggests a more this-worldly diagnosis and prescription than any of the religious traditions considered so far.

These extracts are from the translation by E. R. Hughes, in *Chinese Philosophy in Classical Times* (Everyman's Library: Dent and Dutton, 1942; in the United States, E. P. Dutton).

THE GOODNESS OF HUMAN NATURE, AND THE CONDITIONS FOR IT TO FLOURISH

Master Kao said that that which is born in men is their nature, and when Master Meng asked him whether the meaning of this statement was like the meaning of "white is white," Kao answered that that was his meaning.

Master Kao said, "Men's nature is like a current of water. If you open a channel for the current to the east, it will flow east. If you open a channel to the west, it will flow west. Men's nature makes no distinction between the good and the not good, just as water makes no distinction between east and west." Master Meng replied, "Water can be trusted not to make a distinction between east and west: but is this so in relation to up and down? Men's natural tendency towards goodness is like the water's tendency to find the lower level. Now if, for example, you strike the water and make it leap up, it is possible to force it over your head. . . . But this surely is not the nature of water, and it is only if force is applied that it acts in this way. That men can be made to do evil is due to their nature also being like this.". . .

Kung Tu [a disciple of Mencius] said, "Master Kao says that men's nature is neither good nor evil . . . whilst there are others who say that some men have a good nature and some have an evil nature. . . . Now you [i.e. Mencius] say that men's nature is good. If this is the case, then are all these others wrong?"

Master Meng replied, "Speaking realistically, it is possible for men to be good, and that is what I mean when I say that (men's nature) is good. If they become evil, it is not the fault of their natural powers. Thus all men have a sense of compassion, also a sense of shame over wickedness, a sense of reverence, and a sense of truth and error. The sense of compassion is equivalent to individual morality, the sense of shame to public morality, the sense of reverence to ritual propriety, and the sense of right and wrong equals wisdom. These four, individual morality, public morality, ritual propriety, and wisdom, are not fused into us from without. We invariably are possessed of them, and that without reflecting on them. This is why I maintain that if we seek for them, then we find them: if we neglect them, we lose them. That contrasts can be made between men of twice and five times and even to an incalculable degree is due to the fact that men fail fully to carry out their natural powers. . . ."

Master Meng said, "All men have the sense of compassion for others. The former kings, having this sense of compassion, thereby ruled compassionately. Having the sense of compassion and practising this compassionate rule, their control of the Great Society was as easy as rolling things in the hand. What I mean by all men having a sense of compassion is that if, for instance, a child is suddenly seen to be on the point of falling into a well, everybody without exception will have a sense of distress. It is not by reason of any close intimacy with the parents of the child, nor by reason of a desire for the praise of neighbours and friends, nor by reason of disliking to be known as the kind of man (who is not moved by compassion). From this point of view we observe that it is inhuman to have no sense of compassion, inhuman to have no sense of shame over wickedness, inhuman to have no sense of modesty and the need for yielding place to a better man, inhuman not to distinguish right and wrong. The sense of compassion represents the tender shoot of individual morality, the sense of shame that of public morality, the sense of modesty that of ritual propriety, the sense of truth and error that of wisdom. Men have these four tender shoots just as they have their four limbs; and the man who in spite of having these tender shoots in him says of his own accord 'I am unable,' that man plays the thief with himself; and when he says it of his ruler, he plays the thief of his ruler. . . . Every 'I' with these four tender shoots knows how to nourish and expand them, just like fire bursting into flame and a spring gushing forth on all sides. Let them expand to the full, and they alone suffice to protect all within the Four Seas. Should they be prevented from expansion, they do not suffice for the service of a man's father and mother."

Master Meng said, ". . . With those who do violence to themselves it is impossible to converse, with those who throw themselves away it is impossible to act. The meaning of doing violence to oneself is contravention of ritual and righteousness, the meaning of throwing oneself away is inability in oneself to dwell in human-heartedness and follow righteousness. For human-heartedness is man's abode of peace and righteousness is man's true path. Alas, that that abode is left empty and desolate and that path is abandoned and not followed!"

Kun-sun Ch'ou [a disciple] asked if he might learn from his Master what his idea of an unperturbed mind was and what Master Kao's was. The reply was, "Kao's idea is, do not try to get in the mind what you cannot put into words, do not try to get from the vital energy in you what you cannot get from your mind. Now the second statement is permissible but the first is not. Purpose in the mind is the teaching power (needed by) the vital energy, as this energy is the power developing cohesion (needed by) the body; and of these purpose is of the first importance, the vital energy of secondary importance. The result is that I maintain that we have to hold fast to our purposes, but not if these injure the vital energy in us. . . . If the purposes be integrated, they can stir up the vital energy in us; and (equally) if the vital energy in our limbs be integrated, it can stir up purpose in the mind. For example, when a man stumbles or gets hurried, this is due to (unintegrated) energy in his limbs, and it has a reversing [? paralysing] effect on the mind."

Kung-sun Ch'ou then asked in what way (spiritual) growth was achieved. The reply was, "By our understanding (the significance of) speech, and by skill in nourishing the vast-flowing vital energy in man." When Kung-sun Ch'ou asked him what he meant by "the vast-flowing energy," the reply was, "It is difficult to put it into words. Such is the nature of this energy that it is immensely great and immensely strong, and if it be nourished by uprightness and so sustain no injury, then it pervades the whole space between the heavens and the earth. Such is the nature of this force that it marries righteousness with truth[*Tao*] and without it (material and spiritual) corruption would set in. It is the product of accumulated righteousness, though not of a righteousness handed down and casually caught at. (For) if human conduct be possessed of no (divine) discontent in the mind, then corruption would set in. . . . Here is a duty which must be accomplished, and that without ceasing. The mind must not forget it. And yet the mind must not deliberately help the growth, as the Sung farmer did. There was a man there who was vexed with his growing corn because it was not tall; so he pulled it up. When he returned home in a state of exhaustion he told his people, 'I am very tired to-day: I have been helping the corn to grow.' His son ran out to see—the corn of course was all withered away. Now in our Great Society there are very few who (in relation to the vast-flowing vital energy) either do not help the corn to grow, or neglect it as being of no use. . . . The people who help the vast-flowing vital energy to grow are the people who pull it up. Not only is their labour in vain, it is actually injurious."

Master Meng said, "Those who have the Mean nurture those who have not, and those who have natural gifts nurture those who have not. Thus it is that men are glad over the possession of worthy fathers and elder brothers. If those who had the Mean and natural gifts were to forsake those who had not, the difference between the worthy and unworthy could not amount to the space of an inch."

Master Meng said, ". . . An enlightened man builds on the deep foundation of the Way [*Tao*]. His wish is to possess it of himself. If he comes to possess it, then he dwells at peace in it and so comes to have a profound confidence in it, and so gets it on every hand and makes contact with its bubbling spring. This is the cause of an enlightened man's wishing to possess it of himself."

From the New Testament

The Jewish religious teacher Jesus of Nazareth wrote nothing, or at least nothing we know of. The new religion of Christianity developed out of the reactions of his followers to his life and death, and alleged resurrection. In the letters of St. Paul to the early Christian churches, which predate the writing of the gospel accounts of Jesus' ministry, we find the earliest formulations of the new doctrines of the divinity of Jesus Christ and his God-given power to redeem men from their sinful condition and give them everlasting life. This life is described in terms of the resurrection of a new "spiritual body," not as the survival of a disembodied soul. The opposition between spirit and flesh, which recurs in St. Paul's writing, seems to mean not a Platonic dualism of soul and body, but the difference between redeemed and sin-dominated life.

These extracts are from the Epistle to the Romans (Chapters 7 and 8) and from the First Epistle to the Corinthians (Chapter 15); the translation from the Greek is the Revised Standard Version. In Part Two, we will examine Aquinas' formulation of a Christian doctrine of human nature in terms of Aristotelean philosophy.

THE REDEMPTIVE POWER OF CHRIST

We know that the law is spiritual; but I am carnal, sold under sin. I do not understand my own actions. For I do not do what I want, but I do the very thing I hate.

Now if I do what I do not want, I agree that the law is good. So then it is no longer I that do it, but sin which dwells within me. For I know that nothing good dwells within me, that is, in my flesh. I can will what is right, but I cannot do it. For I do not do the good I want, but the evil I do not want is what I do. Now if I do what I do not want, it is no longer I that do it, but sin which dwells within me.

So I find it to be a law that when I want to do right, evil lies close at hand. For I delight in the law of God, in my inmost self, but I see in my members another law at war with the law of my mind and making me captive to the law of sin which dwells in my members. Wretched man that I am! Who will deliver me from this body of death? Thanks be to God through Jesus Christ our Lord! So then, I of myself serve the law of God with my mind, but with my flesh I serve the law of sin.

There is therefore now no condemnation for those who are in Christ Jesus. For the law of the Spirit of life in Christ Jesus has set me free from the law of sin and death. For God has done what the law, weakened by the flesh, could not do: sending his own Son in the likeness of sinful flesh and for sin, he condemned sin in the flesh, in order that the just requirement of the law might be fulfilled in us, who walk not according to the flesh but according to the Spirit. For those who live according to the flesh set their minds on the things of the flesh, but those who live according to the Spirit set their minds on the things of the Spirit. To set the mind on the flesh is death, but to set the mind on the Spirit is life and peace. For the mind that is set on the flesh is hostile to God; it does not submit to God's law, indeed it cannot; and those who are in the flesh cannot please God.

But you are not in the flesh, you are in the Spirit, if in fact the Spirit of God dwells in you.

THE RESURRECTION AND THE LIFE TO COME

Now I would remind you, brethren, in what terms I preached to you the gospel, which you received, in which you stand, by which you are saved, if you hold it fast—unless you believed in vain.

For I delivered to you as of first importance what I also received, that Christ died for our sins in accordance with the scriptures, that he was buried, that he was raised on the third day in accordance with the scriptures, and that he appeared to Cephas, then to the twelve. Then he appeared to more than five hundred brethren at one time, most of whom are still alive, though some have fallen asleep. Then he appeared to James, then to all the apostles. Last of all, as to one untimely born, he appeared also to me. . . .

Now if Christ is preached as raised from the dead, how can some of you say that there is no resurrection of the dead? But if there is no resurrection of the dead, then Christ has not been raised; if Christ has not been raised, then our preaching is in vain and your faith is in vain. We are even found to be misrepresenting God, because we testified of God that he raised Christ, whom he did not raise if it is true that the dead are not raised.

For if the dead are not raised, then Christ has not been raised. If Christ has not been raised, your faith is futile and you are still in your sins.

Then those also who have fallen asleep in Christ have perished. If for this life only we have hoped in Christ, we are of all men most to be pitied.

But in fact Christ has been raised from the dead, the first fruits of those who have fallen asleep. For as by a man came death, by a man has come also the resurrection of the dead. For as in Adam all die, so also in Christ shall all be made alive. . . .

But some one will ask, "How are the dead raised? With what kind of body do they come?" You foolish man! What you sow does not come to life unless it dies. And what you sow is not the body which is to be, but a bare kernel, perhaps of wheat or of some other grain. But God gives it a body as he has chosen, and to each kind of seed its own body.

For not all flesh is alike, but there is one kind for men, another for animals, another for birds, and another for fish.

There are celestial bodies and there are terrestrial bodies; but the glory of the celestial is one, and the glory of the terrestrial is another. There is one glory of the sun, and another glory of the moon, and another glory of the stars; for star differs from star in glory.

So is it with the resurrection of the dead. What is sown is perishable, what is raised is imperishable. It is sown in dishonor, it is raised in glory. It is sown in weakness, it is raised in power. It is sown a physical body, it is raised a spiritual body. If there is a physical body, there is also a spiritual body. Thus it is written, "The first man Adam became a living being"; the last Adam became a life-giving spirit. But it is not the spiritual which is first but the physical, and then the spiritual. The first man was from the earth, a man of dust; the second man is from heaven. As was the man of dust, so are those who are of the dust; and as is the man of heaven, so are those who are of heaven. Just as we have borne the image of the man of dust, we shall also bear the image of the man of heaven. I tell you this, brethren: flesh and blood cannot inherit the kingdom of God, nor does the perishable inherit the imperishable.

Lo! I tell you a mystery. We shall not all sleep, but we shall all be changed, in a moment, in the twinkling of an eye, at the last trumpet. For the trumpet will sound, and the dead will be raised imperishable, and we shall be changed. For this perishable nature must put on the imperishable, and this mortal nature must put on immortality. When the perishable puts on the imperishable, and the mortal puts on immortality, then shall come to pass the saying that is written:

> "Death is swallowed up in victory."
> "O death, where is thy victory?
> O death, where is thy sting?"

The sting of death is sin, and the power of sin is the law. But thanks be to God, who gives us the victory through our Lord Jesus Christ.

Part II

Reasoned Argument about Human Nature in Greek and Medieval Philosophy

What distinguishes the following readings as *philosophy,* as opposed to the religious writings sampled so far, is their attempt to give rational *argument* for the propositions asserted. There is more than just a claim to knowledge of the truth, some account is offered of *how* the truths are supposed to be known, and hence of what reason there is to believe them. (Even the cleverness of the Venerable Nagasena was devoted to answering King Milinda's objections, rather than to giving any direct proof of Buddhist doctrine.)

As far as we know, the ancient Greeks were the first thus to argue. We find in them the root of the whole Western tradition of systematic enquiry: the assumption that, by careful exercise of our reasoning abilities, we can find out new truths about the universe and our place within it. Socrates is the patron saint of philosophy, in virtue of his practice of rational argument and criticism and his martyrdom for the cause of free enquiry after the truth.

In the Middle Ages there was a synthesis of Greek rationality and Christian faith. But the distinction between philosophy and theology (or, within the latter, between natural and revealed theology) was made in terms of the method of proof—as an appeal either to unaided human reason, or to Divine authority as expressed in the Scriptures or in the Church. Philosophical argument is therefore present within the writings of the Age of Faith, above all in the great system of Aquinas.

Plato

If Socrates is the patron saint of philosophy, Plato (about 430–347 B.C.) is the first to have left behind substantial philosophical writings to influence posterity. Most of these are cast in the form of dialogues in which the leading part is played by Socrates, who was Plato's teacher. The style ranges from the vivid imagery of poetry and myth to the dry technicalities of logical argument.

The first excerpt here is from the *Phaedo* (77–80), where Plato takes as already proved his theory of absolute unchanging Forms and his belief in the existence of the soul before birth. He then goes on to argue that the disembodied soul survives to enter into a better world after death.

Next, we turn to the famous passage in the *Republic* (514–519) in which Plato likens the unenlightened human condition to that of prisoners chained in a cave, and speculates on how some may gain knowledge of the realities outside the cave and be induced to apply their knowledge for the benefit of the rest of humanity.

In the third selection, from the *Republic* (369, 434–444), Plato presents an elaborate analogy between what he calls "justice" in the state and "justice" in the individual. Having already argued for a division of the state into three classes of guardians, auxiliaries, and traders, he now proceeds to his famous tripartite theory of the soul, dividing it into reason, passion or spirit, and appetite or desire. "Justice," which amounts here to well-being or health, is said to consist in the harmonious combination of the different elements, with reason as ruler of the individual soul, and the educated guardians exercising political power in the state.

Finally, a brief example of Plato's more colourful poetic or religious style is exhibited in an excerpt from the *Phaedrus* (253–254), where he illustrates his tripartite theory of human nature by the image of the charioteer pulled apart by his good and bad horses. The internal conflicts thus memorably pictured are described in strikingly parallel terms by Freud.

The translation is by Benjamin Jowett (Oxford University Press, 1st ed., 1871, 4th ed., 1953).

THE IMMORTALITY OF THE SOUL

Then may we not say, Simmias, that if there do exist these things of which we are always talking, absolute beauty and goodness, and all that class of realities; and if to this we refer all our sensations and with this compare them, finding the realities to be pre-existent and our own possession—then just as surely as these exist, so surely must our souls have existed before our birth? Otherwise our whole argument would be worthless. By an equal compulsion we must believe both that these realities exist, and that our souls existed before our birth; and if not the realities, then not the souls.

Yes, Socrates; I am convinced that there is precisely the same necessity for the one as for the other; and the argument finds a safe refuge in the position that the existence of the soul before birth cannot be separated from the existence of the reality of which you speak. For there is nothing which to my mind is so patent as that beauty, goodness, and the other realities of which you were just now speaking, exist in the fullest possible measure; and I am satisfied with the proof.

Well, but is Cebes satisfied? for I must convince him too.

I think, said Simmias, that Cebes is satisfied: although he is the most incredulous of mortals, yet I believe that he is sufficiently convinced of the existence of the soul before birth. But that after death the soul will continue to exist is not yet proven even to my own satisfaction. I cannot get rid of the objection to which Cebes was referring—the common fear that at the moment when the man dies the soul is dispersed, and that this may be the end of her. For admitting that she may have come into being and been framed out of some unknown other elements, and was in existence before entering the human body, why after having entered in and gone out again may she not herself be destroyed and come to an end?

Very true, Simmias, said Cebes; it appears that about half of what was required has been proven; to wit, that our souls existed before we were born;—that the soul will exist after death as well as before birth is the other half of which the proof is still wanting, and has to be supplied; when that is given the demonstration will be complete.

But that proof, Simmias and Cebes, has been already given, said Socrates, if you put the two arguments together—I mean this and the former one, in which we agreed that everything living is born of the dead. For if the soul exists before birth, and in coming to life and being born can be born only from death and the state of death, must she not after death continue to exist, since she has to be born again?—Surely the proof which you desire has been already furnished. Still I suspect that you and Simmias would be glad to probe the argument further. Like children, you are haunted with a fear that when the soul leaves the body, the wind may really blow her away and scatter her; especially if a man should happen to die in a great storm and not when the weather is calm.

Cebes answered with a smile: Then, Socrates, you must argue us out of our fears—and yet, strictly speaking, they are not our fears, but perhaps even in us men there is a child to whom death is a sort of hobgoblin: him too we must persuade not to be afraid.

Socrates said: Let the voice of the charmer be applied daily until you have charmed away the fear.

And where shall we find a good charmer of our fears, Socrates, now that you are abandoning us?

Hellas, he replied, is a large place, Cebes, and has good men, and there are barbarous races not a few: seek for him among them all, far and wide, sparing neither pains nor money; for there is no better way of spending your money. And you must seek yourselves too, along with one another; for perhaps you will not easily find others better able to do it.

The search, replied Cebes, shall certainly be made. And now, if you please, let us return to the point of the argument at which we digressed.

By all means, replied Socrates; what else should I please?

Very good.

Must we not, said Socrates, ask ourselves what kind of thing that is which is liable to be scattered, and for what kind of thing we ought to fear that fate? and what is that for which we need have no fear? And then we may proceed further to inquire to which of the two classes soul belongs—our hopes and fears as to our own souls will turn upon the answers to these questions.

Very true, he said.

Now that which is compounded and is by nature composite may be supposed to be therefore capable, as of being compounded, so also of being dissolved; but that which is not composite, and that only, must be, if anything is, indissoluble.

Yes; I should imagine so, said Cebes.

And the non-composite may be assumed to be the same and unchanging, whereas the composite is always changing and never the same.

I agree, he said.

Then now let us return to the previous discussion. Is that reality of whose being we give account in the dialectical process—whether equality, beauty, or anything else—are these realities, I say, liable at times to some degree of change? or are they each of them always what they are, having the same uniform self-existent and unchanging natures, not admitting of variation at all, or in any way, or at any time?

They must be always the same, Socrates, replied Cebes.

And what would you say of the many beautiful, for instance, men or horses or garments or any other such things, or of the many equal, or generally of all the things which are named by the same names as the realities—are they the same always? May they not rather be described in exactly opposite terms, as almost always changing and hardly ever the same either with themselves or with one another?

The latter, replied Cebes; they are always in a state of change.

And these you can touch and see and perceive with the senses, but the unchanging things you can only grasp with the mind—they are invisible and are not seen?

That is very true, he said.

Well then, added Socrates, let us suppose that there are two sorts of existences—one seen, the other unseen.

Let us suppose them.

The seen is the changing, and the unseen is the unchanging?

That may be also supposed.

And, further, of ourselves is not one part body, another part soul?

To be sure.

And to which class is the body more alike and akin?

Clearly to the seen—no one can doubt that.

And is the soul seen or not seen?

Not by man, Socrates.

And what we mean by 'seen' and 'not seen' is that which is or is not visible to the eye of man?

Yes, to the eye of man.

And is the soul seen or not seen?

Not seen.

Unseen then?

Yes.

Then the soul is more like to the unseen, and the body to the seen?

That follows necessarily, Socrates.

And were we not saying some time ago that the soul when using the body as an instrument of perception, that is to say, when using the sense of sight or hearing or some other sense (for the meaning of perceiving through the body is perceiving through the senses)—were we not saying that the soul too is then dragged by the body into the region of the changeable, and wanders and is confused; the world spins round her, and she is like a drunkard, when she touches change?

Very true.

But when returning into herself she reflects, then she passes into the other world, the region of purity, and eternity, and immortality, and unchangeableness, which are her kindred, and with them she ever lives, when she is by herself and is not let or hindered; then she ceases from her wandering, and being in contact with things unchanging is unchanging in relation to them. And this state of the soul is called wisdom?

That is well and truly said, Socrates, he replied.

And to which class is the soul more nearly alike and akin, as far as may be inferred from this argument, as well as from the preceding one?

I think, Socrates, that, in the opinion of everyone who follows the argument, the soul will be infinitely more like the unchangeable—even the most stupid person will not deny that.

And the body is more like the changing?

Yes.

Yet once more consider the matter in another light: When the soul and the body are united, then nature orders the soul to rule and govern, and the body to obey and serve. Now which of these two functions is like to the divine? and which to the mortal? Does not the divine appear to you to be that which is formed to govern and command, and the mortal to that which is by its nature subject and servant?

True.

And which does the soul resemble?

The soul resembles the divine, and the body the mortal—there can be no doubt of that, Socrates.

Then reflect, Cebes: of all which has been said is not this the conclusion?—that the soul is in the very likeness of the divine, and immortal, and rational, and uniform, and indissoluble, and unchangeable; and that the body is in the very likeness of the human, and mortal, and irrational, and multiform, and dissoluble, and changeable. Can we, my dear Cebes, find any possible ground for rejecting this conclusion?

We cannot.

But if it be true, then is not the body liable to speedy dissolution? and is not the soul almost or altogether indissoluble?

Certainly.

And do you further observe, that after a man is dead, the body, or visible part of him, which is lying in the visible world, and is called a corpse, and would naturally be dissolved and decomposed and dissipated, is not dissolved or decomposed at once, but may remain for some time, nay even for a long time, if the constitution be sound at the time of death, and the season of the year favourable? For the body when shrunk and embalmed, as the manner is in Egypt, may remain almost entire for a prodigious time; and even in decay, there are still some portions, such as the bones and ligaments, which are practically indestructible:—Do you agree?

Yes.

And is it likely that the soul, which is invisible, in passing to the place of the true Hades, which like her is invisible, and pure, and noble, and on her way to the good and wise God, whither, if God will, my soul is also soon to go—that the soul, I repeat, if this be her nature, is blown away and destroyed immediately on quitting the body, as the many say? That can never be, my dear Simmias and Cebes. The truth rather is that the soul which is pure at departing and draws after her no bodily taint, having never voluntarily during life had connexion with the body, which she is ever avoiding, herself gathered into herself, and making such abstraction her perpetual study—all this means that she has been a true disciple of philosophy; and therefore has in fact been always practising how to die without complaint. For is not such a life the practice of death?

Certainly.

That soul, I say, herself invisible, departs to the invisible world—to the divine and immortal and rational: thither arriving, she is secure of bliss and is released from the error and folly of men, their fears and wild passions and all other human ills. . . .

THE CAVE, AND THE WAY OUT OF IT

And now, I said, let me show in a figure how far our nature is enlightened or unenlightened:—Behold! human beings housed in an underground cave, which has a long entrance open towards the light and as wide as the interior of the cave; here they have been from their childhood, and have their legs and necks chained, so that they cannot move and can only see before them, being prevented by the chains from turning round their heads. Above and behind them a fire is blazing at a distance, and between the fire and the prisoners there is a raised way; and you will see, if you look, a low wall built along the way, like the screen which marionette players have in front of them, over which they show the puppets.

I see.

And do you see, I said, men passing along the wall carrying all sorts of vessels, and statues and figures of animals made of wood and stone and various materials, which appear over the wall? While carrying their burdens, some of them, as you would expect, are talking, others silent.

You have shown me a strange image, and they are strange prisoners.

Like ourselves, I replied; for in the first place do you think they have seen anything of themselves, and of one another, except the shadows which the fire throws on the opposite wall of the cave?

How could they do so, he asked, if throughout their lives they were never allowed to move their heads?

And of the objects which are being carried in like manner they would only see the shadows?

Yes, he said.

And if they were able to converse with one another, would they not suppose that the things they saw were the real things?

Very true.

And suppose further that the prison had an echo which came from the other side, would they not be sure to fancy when one of the passers-by spoke that the voice which they heard came from the passing shadow?

No question, he replied.

To them, I said, the truth would be literally nothing but the shadows of the images.

That is certain.

And now look again, and see in what manner they would be released from their bonds, and cured of their error, whether the process would naturally be as follows. At first, when any of them is liberated and compelled suddenly to stand up and turn his neck round and walk and look towards the light, he will suffer sharp pains; the glare will distress him, and he will be unable to see the realities of which in his former state he had seen the shadows; and then conceive someone saying to him that what he saw before was an illusion but that now, when he is approaching nearer to being and his eye is turned towards more real existence, he has a clearer vision,—what will be his reply? And you may further imagine that his instructor is pointing to the objects as they pass and requiring him to name them,—will he not be perplexed? Will he not fancy that the shadows which he formerly saw are truer than the objects which are now shown to him?

Far truer.

And if he is compelled to look straight at the light, will he not have a pain in his eyes which will make him turn away to take refuge in the objects of vision which he can see, and which he will conceive to be in reality clearer than the things which are now being shown to him?

True, he said.

And suppose once more, that he is reluctantly dragged up that steep and rugged ascent, and held fast until he is forced into the presence of the sun himself, is he not likely to be pained and irritated? When he approaches the light his eyes will be dazzled, and he will not be able to see anything at all of what are now called realities.

Not all in a moment, he said.

He will require to grow accustomed to the sight of the upper world. And first he will see the shadows best, next the reflections of men and other objects in the water, and then the objects themselves; and, when he turned to the heavenly bodies and the heaven itself, he would find it easier to gaze upon the light of the moon and the stars at night than to see the sun or the light of the sun by day?

Certainly.

Last of all he will be able to see the sun, not turning aside to the illusory reflections of him in the water, but gazing directly at him in his own proper place, and contemplating him as he is.

Certainly.

He will then proceed to argue that this is he who gives the seasons and the years, and is the guardian of all that is in the visible world, and in a certain way the cause of all things which he and his fellows have been accustomed to behold?

Clearly, he said, he would arrive at this conclusion after what he had seen.

And when he remembered his old habitation, and the wisdom of the cave and his fellow-prisoners, do you not suppose that he would felicitate himself on the change, and pity them?

Certainly, he would.

And if they were in the habit of conferring honours among themselves on those who were quickest to observe the passing shadows and to remark which of them went before and which followed after and which were together, and who were best able from these observations to divine the future, do you think that he would be eager for such honours and glories, or envy those who attained honour and sovereignty among those men? Would he not say with Homer,

"Better to be a serf, labouring for a landless master,"

and to endure anything, rather than think as they do and live after their manner?

Yes, he said, I think that he would consent to suffer anything rather than live in this miserable manner.

Imagine once more, I said, such a one coming down suddenly out of the sunlight, and being replaced in his old seat; would he not be certain to have his eyes full of darkness?

To be sure, he said.

And if there were a contest, and he had to compete in measuring the shadows with the prisoners who had never moved out of the cave, while his sight was still weak, and before his eyes had become steady (and the time which would be needed to acquire this new habit of sight might be very considerable), would he not make himself ridiculous? Men would say of him that he had returned from the place above with his eyes ruined; and that it was better not even to think of ascending; and if anyone tried to loose another and lead him up to the light, let them only catch the offender, and they would put him to death.

No question, he said.

This entire allegory, I said, you may now append, dear Glaucon, to the previous argument; the prison-house is the world of sight, the light of the fire is the power of the sun, and you will not misapprehend me if you interpret the journey upwards to be the ascent of the soul into the intellectual world according to my surmise, which, at your desire, I have expressed—whether rightly or wrongly God knows. But, whether true or false, my opinion is that in the world of knowledge the Idea of good appears last of all, and is seen only with an effort; although, when seen, it is inferred to be the universal author of all things beautiful and right, parent of light and of the lord of light in the visible world, and the immediate and supreme source of reason and truth in the intellectual; and that this is the power upon which he who would act rationally either in public or private life must have his eye fixed.

I agree, he said, as far as I am able to understand you.

Moreover, I said, you must agree once more, and not wonder that those who attain to this vision are unwilling to take any part in human affairs; for their souls are ever hastening into the upper world where they desire to dwell; which desire of theirs is very natural, if our allegory may be trusted.

Yes, very natural. . . .

Yes, I said; and there is another thing which is likely, or rather a necessary inference from what has preceded, that neither the uneducated and uninformed of the truth, nor yet those who are suffered to prolong their education without end, will be able ministers of State; not the former, because they have no single aim of duty which is the rule of all their actions, private as well as public; nor the latter, because they will not act at all except upon compulsion, fancying that they are already dwelling apart in the islands of the blest.

Very true, he replied.

Then, I said, the business of us who are the founders of the State will be to compel the best minds to attain that knowledge which we have already shown to be the greatest of all, namely, the vision of the good; they must make the ascent which we have described; but when they have ascended and seen enough we must not allow them to do as they do now.

What do you mean?

They are permitted to remain in the upper world, refusing to descend again among the prisoners in the cave, and partake of their labours and honours, whether they are worth having or not.

But is not this unjust? he said; ought we to give them a worse life, when they might have a better?

You have again forgotten, my friend, I said, the intention of our law, which does not aim at making any one class in the State happy above the rest; it seeks rather to spread happiness over the whole State, and to hold the citizens together by persuasion and necessity, making each share with others any benefit which he can confer upon the State; and the law aims at producing such citizens, not that they may be left to please themselves, but that they may serve in binding the State together.

True, he said, I had forgotten.

Observe, Glaucon, that we shall do no wrong to our philosophers but rather make a just demand, when we oblige them to have a care and providence of others; we shall explain to them that in other States, men of their class are not obliged to share in the toils of politics: and this is reasonable, for they grow up spontaneously, against the will of the governments in their several States; and things which grow up of themselves, and are indebted to no one for their nurture, cannot fairly be expected to pay dues for a culture which they have never received. But we have brought you into the world to be rulers of the hive, kings of yourselves and of the other citizens, and have educated you far better and more perfectly than they have been educated and you are better able to share in the double duty. Wherefore each of you, when his turn comes, must go down to rejoin his companions, and acquire with them the habit of seeing things in the dark. As you acquire that habit, you will see ten thousand times better than the inhabitants of the cave, and you will know what the several images are and what they represent, because you have seen the beautiful and just and good in their truth. And thus our State, which is also yours, will be a reality and not a dream only, and will be administered in a spirit unlike that of other States, in which men fight with one another about shadows only and are distracted in the struggle for power, which in their eyes is a great good. Whereas the truth is that the State in which those who are to govern have least ambition to do so is always the best and most quietly governed, and the State in which they are most eager, the worst.

JUSTICE IN SOCIETY,
AND THE TRIPARTITE STRUCTURE OF THE SOUL

. . . [J]ustice is, as you know, sometimes spoken of as the virtue of an individual, and sometimes as the virtue of a State.

True, he replied.

And is not a State larger than an individual?

It is.

Then in the larger, justice is likely to be more abundant and more easily discernible. I propose therefore that we inquire into the nature of justice and injustice, first as they appear in the State, and secondly in the individual, proceeding from the greater to the lesser and comparing them.

That, he said, is an excellent proposal.

And if we imagine the State in process of creation, we shall see the justice and injustice of the State in process of creation also.

I dare say.

When the State is completed there may be a hope that the object of our search will be more easily discovered.

Yes, far more easily.

But ought we to attempt to construct one? I said; for to do so, as I am inclined to think, will be a very serious task. Reflect therefore.

I have reflected, said Adeimantus, and am anxious that you should proceed.

A State, I said, arises, as I conceive, out of the needs of mankind; no one is self-sufficing, but all of us have many wants. Can any other origin of a State be imagined?

There can be no other.

Then, as we have many wants, and many persons are needed to supply them, one takes a helper for one purpose and another for another; and when these partners and helpers are gathered together in one habitation the body of inhabitants is termed a State.

True, he said.

And it is in the belief that it is for his own good, that one man gives to another or receives from him in exchange.

Very true.

Then, I said, let us construct a State in theory from the beginning. . . .

Seeing then, I said, that there are three distinct classes, any meddling of one with another, or the change of one into another, is the greatest harm to the State, and may be most justly termed evil-doing?

Precisely.

And the greatest degree of evil-doing to one's own city would be termed by you injustice?

Certainly.

This then is injustice; and on the other hand when the three main classes, traders, auxiliaries, and guardians, each do their own business, that is justice, and will make the city just.

I agree with you.

We will not, I said, be over-positive as yet; but if, on trial, this conception of justice be verified in the individual as well as in the State, there will be no longer any room for doubt; if it be not verified, we must have a fresh inquiry. First let us complete the old investigation, which we began, as you remember, under the impression that, if we could previously examine justice on the larger scale, there would be less difficulty in discerning her in the individual. That larger example appeared to be the State, and accordingly we constructed as good a one as we could, knowing well that in the good State justice would be found. Let the discovery which we made be now applied to the individual—if they agree, we shall be satisfied; or, if there be a difference in the individual, we will come back to the State and have another trial of the theory. The friction of the two when rubbed together may possibly strike the light of justice, from which we can kindle a steady flame in our souls.

That will be in regular course; let us do as you say.

I proceeded to ask: When two things, a greater and less, are called by the same name, are they like or unlike in so far as they are called the same?

Like, he replied.

The just man then, if we regard the idea of justice only, will be like the just State?

He will.

And a State was thought by us to be just when the three classes in the State severally did their own business; and also thought to be temperate and valiant and wise by reason of certain other affections and qualities of these same classes?

True, he said.

And so of the individual: we may assume that he has the same three principles in his own soul which are found in the State; and he may be rightly described in the same terms, because he is affected in the same manner?

Certainly, he said.

Once more then, O my friend, We have alighted upon an easy question—whether the soul has these three principles or not? . . .

But the question is not quite so easy when we proceed to ask whether these principles are three or one; whether, that is to say, we learn with one part of our nature, are angry with another, and with a third part desire the satisfaction of our natural appetites; or whether the whole soul comes into play in each sort of action—to determine that is the difficulty.

Yes, he said; there lies the difficulty.

Then let us now try and determine whether they are the same or different.

How?

Clearly the same thing cannot act or be acted upon in the same part or in relation to the same thing at the same time, in contrary ways; and therefore whenever this contradiction occurs in things apparently the same, we know that they are really not the same, but different.

Good.

For example, I said, can the same thing be at rest and in motion at the same time in the same part?

Impossible.

Now, I said, let us have still more precise understanding, lest we should hereafter fall out by the way. Imagine the case of a man who is standing and also moving his hands and his head, and suppose a person to say that one and the same person is in motion and at rest at the same moment—to such a mode of speech we should object, and should rather say that one part of him is in motion while another is at rest.

Very true. . . .

Well, I said, would you not allow that assent and dissent, desire and aversion, attraction and repulsion, are all of them opposites, whether they are regarded as active or passive (for that makes no difference in the fact of their opposition)?

Yes, he said, they are opposites.

Well, I said, and hunger and thirst, and the desires in general, and again willing and wishing,—all these you would refer to the classes already mentioned. You would say—would you not?—that the soul of him who desires is either seeking after the object of desire; or is drawing towards herself the thing which she wishes to possess: or again,—for she may merely consent that something should be offered to her—intimates her wish to have it by a nod of assent, as if she had been asked a question?

Very true. . . .

Then the soul of the thirsty one, in so far as he is thirsty, desires only drink; for this she yearns, and for this she strives?

That is plain.

And if you suppose something which pulls a thirsty soul away from drink that must be different from the thirsty principle which draws him like a beast to drink; for, as we were saying, the same thing cannot at the same time with the same part of itself act in contrary ways about the same.

Impossible.

No more than you can say that the hands of the archer push and pull the bow at the same time, but what you say is that one hand pushes and the other pulls.

Exactly so, he replied.

Now are there times when men are thirsty, and yet unwilling to drink?

Yes, he said, it constantly happens.

And in such a case what is one to say? Would you not say that there was something in the soul bidding a man to drink and something forbidding him, which is other and stronger than the principle which bids him?

I should say so.

And the prohibition in such cases is derived from reasoning, whereas the motives which lead and attract proceed from passions and diseases?

Clearly.

Then we may fairly assume that they are two, and that they differ from one another; the one with which a man reasons, we may call the rational principle of the soul, the other, with which he loves and hungers and thirsts and feels the flutterings of any other desire, may be termed the irrational or appetitive, the ally of sundry pleasures and satisfactions?

Yes, he said, we may fairly assume them to be different.

So much, then, for the definition of two of the principles existing in the soul. And what now of passion, or spirit? Is it a third, or akin to one of the preceding?

I should be inclined to say—akin to desire.

Well, I said, there is a story which I remember to have heard, and in which I put faith. The story is, that Leontius, the son of

Aglaion, coming up one day from the Piraeus, under the north wall on the outside, observed some dead bodies lying on the ground at the place of execution. He felt a desire to see them, and also a dread and abhorrence of them; for a time he struggled and covered his eyes, but at length the desire got the better of him; and forcing them open, he ran up to the dead bodies, saying, Look, ye wretches, take your fill of the fair sight.

I have heard the story myself, he said.

The moral of the tale is that anger at times goes to war with desire, as though they were two distinct things.

Yes; that is the meaning, he said.

And are there not many other cases in which we observe that when a man's desires violently prevail over his reason, he reviles himself, and is angry at the violence within him, and that in this struggle, which is like the struggle of factions in a State, his spirit is on the side of his reason;—but for the passionate or spirited element to take part with the desires when reason decides that she should not be opposed, is a sort of thing which I believe that you never observed occurring in yourself, nor, as I should imagine, in anyone else?

Certainly not.

Suppose that a man thinks he has done a wrong to another, the nobler he is the less able is he to feel indignant at any suffering, such as hunger, or cold, or any other pain which the injured person may inflict upon him—these he deems to be just, and, as I say, his spirit refuses to be excited by them.

True, he said.

But when a man thinks that he is the sufferer of the wrong, then the spirit within him boils and chafes, and is on the side of what it believes to be justice; and though it suffers hunger or cold or other pain, it is only the more determined to persevere and conquer. Such a noble spirit will not be quelled until it has achieved its object or been slain, or until it has been recalled by the reason within, like a dog by the shepherd?

The illustration is perfect, he replied; and in our State, as we were saying, the auxiliaries were to be dogs, and to hear the voice of the rulers, who are their shepherds.

Yes, I said, you understand me admirably; there is, however, a further point which I wish you to consider.

What point?

You remember that passion or spirit appeared at first sight to be a kind of desire, but now we should say quite the contrary; for in the conflict of the soul spirit is arrayed on the side of the rational principle.

Most assuredly.

But a further question arises: Is passion different from reason also, or only a kind of reason; in which latter case, instead of three principles in the soul, there will only be two, the rational and the concupiscent? or rather, as the State was composed of three classes, traders, auxiliaries, counsellors, so may there not be in the individual soul a third element which is passion or spirit, and when not corrupted by bad education is the natural auxiliary of reason?

Yes, he said, there must be a third.

Yes, I replied, if passion, which has already been shown to be different from desire, turn out also to be different from reason.

But that is easily proved:—We may observe even in young children that they are full of spirit almost as soon as they are born, whereas some of them never seem to attain to the use of reason, and most of them late enough.

Excellent, I said, and you may see passion equally in brute animals, which is a further proof of the truth of what you are saying. And we may once more appeal to the words of Homer, which have been already quoted by us,

"He smote his breast, and thus rebuked his heart;"

for in this verse Homer has clearly supposed the power which reasons about the better and worse to be different from the unreasoning anger which is rebuked by it.

Very true, he said.

And so, after much tossing we have reached land, and are fairly agreed that the same principles which exist in the State exist also in the individual, and that they are three in number.

Exactly.

Must we not then infer that the individual is wise in the same way and in virtue of the same quality which makes the State wise?

Certainly.

Also that the State is brave in the same way and by the same quality as an individual is brave, and that there is the same correspondence in regard to the other virtues?

Assuredly.

Therefore the individual will be acknowledged by us to be just in the same way in which the State has been found just?

That follows of course.

We cannot but remember that the justice of the State consisted in each of the three classes doing the work of its own class?

I do not think we have forgotten, he said.

We must now record in our memory that the individual in whom the several components of his nature do their own work will be just, and will do his own work?

Yes, he said, we must record that important fact.

First, it is proper for the rational principle, which is wise, and has the care of the whole soul, to rule, and for the spirit to be the subject and ally?

Certainly.

And, as we were saying, the blending of music and gymnastic will bring them into accord, nerving and sustaining the reason with noble words and lessons, and moderating and soothing and civilizing the wildness of passion by harmony and rhythm?

Quite true, he said.

And these two, thus nurtured and educated, and having learned truly to know their own functions, will rule over the concupiscent, which in each of us is the largest part of the soul and by nature most insatiable of gain; over this they will keep guard, lest, waxing great and strong with the fullness of bodily pleasures, as they are termed, the concupiscent soul, no longer confined to her own sphere, should attempt to enslave and rule those who are not her natural-born subjects, and overturn the whole life of man?

Very true, he said.

Both together will they not be the best defenders of the whole soul and the whole body against attacks from without; the one counselling, and the other going out to fight as the leader directs, and courageously executing his commands and counsels?

True.

Likewise it is by reference to spirit that an individual man is deemed courageous, because his spirit retains in pleasure and in pain the commands of reason about what he ought or ought not to fear?

Right, he replied.

And we call him wise on account of that little part which rules, and which proclaims these commands; the part in which is situated the knowledge of what is for the interest of each of the three parts and of the whole?

Assuredly.

And would you not say that he is temperate who has these same elements in friendly harmony, in whom the one ruling principle of reason, and the two subject ones of spirit and desire, are equally agreed that reason ought to rule, and do not rebel?

Certainly, he said, that is a precise account of temperance whether in the State or individual.

And, finally, I said, a man will be just in that way and by that quality which we have often mentioned.

That is very certain.

And is justice dimmer in the individual, and is her form different, or is she the same which we found her to be in the State?

There is no difference in my opinion, he said.

Because, if any doubt is still lingering in our minds, a few commonplace instances will satisfy us of the truth of what I am saying.

What sort of instances do you mean?

If the case is put to us, must we not admit that the just State, or the man of similar nature who has been trained in the principles of such a State, will be less likely than the unjust to make away with a deposit of gold or silver? Would any one deny this?

No one, he replied.

Will such a man ever be involved in sacrilege or theft, or treachery either to his friends or to his country?

Never.

Neither will he ever, for any reason, break faith where there have been oaths or agreements?

Impossible.

No one will be less likely to commit adultery, neglect his father and mother, or fail in his religious duties?

No one.

And the reason for all this is that each part of him is doing its own business, whether in ruling or being ruled?

Exactly so.

Are you satisfied then that the quality which makes such men and such states is justice, or do you hope to discover some other?

Not I, indeed.

Then our dream has been realized, and the suspicion which we expressed that, at the beginning of our work of construction, some divine power must have conducted us to a primary form of justice, has now been verified?

Yes, certainly.

And the division of labour which required the carpenter and the shoemaker and the rest of them to devote himself to the work for which he is naturally fitted, and to do nothing else, was a shadow of justice, and for that reason it was of use?

Clearly.

And in reality justice was such as we were describing, being concerned however, not with a man's external affairs, but with an inner relationship in which he himself is more truly concerned; for the just man does not permit the several elements within him to interfere with one another, or any of them to do the work of others,—he sets in order his own inner life, and is his own master and his own law, and at peace with himself; and when he has bound together the three principles within him, which may be compared to the higher, lower, and middle notes of the scale, and any that are intermediate between them—when he has bound all these together, and is no longer many, but has become one entirely temperate and perfectly adjusted nature, then he proceeds to act, if he has to act, whether in a matter of property or in the treatment of the body, or in some affair of politics or private business; always thinking and calling that which preserves and co-operates with this harmonious condition, just and good action, and the knowledge which presides over it, wisdom, and that which at any time impairs this condition, he will call unjust action, and the opinion which presides over it ignorance.

You have said the exact truth, Socrates.

Very good; and if we were to affirm that we had discovered the just man and the just State, and the nature of justice in each of them, we should not be far from the truth?

Most certainly not.

May we say so, then?

Let us say so.

And now, I said, injustice has to be considered.

Clearly.

Must not injustice be a strife which arises among the same three principles—a meddlesomeness, and interference, and rising up of a part of the soul against the whole, an assertion of unlawful authority, which is made by a rebellious subject against a true prince, of whom he is the natural vassal,—what is all this confusion and delusion but injustice and intemperance and cowardice and ignorance, and, in short, every form of vice?

Exactly so.

And if the nature of justice and injustice is known, then the meaning of acting unjustly and being unjust, or again of acting justly, is now also perfectly clear?

How so? he said.

Why, I said, they are like disease and health; being in the soul just what disease and health are in the body.

THE WHITE AND DARK HORSES OF THE SOUL

At the beginning of this tale, I divided each soul into three parts—two having the form of horses and the third being like a charioteer; the division may remain. I have said that one horse was good, the other bad, but I have not yet explained in what the goodness or badness of either consists, and to that I will now proceed. The right-hand horse is upright and cleanly made; he has a lofty neck and an aquiline nose; his colour is white, and his eyes dark; he is one who loves honour with modesty and temperance, and the follower of true opinion; he needs no touch of the whip, but is guided by word and admonition only. The other is a crooked, lumbering animal, put together anyhow; he has a short, thick neck; he is flat-faced and of a dark colour, with grey eyes and blood-red complexion; the mate of insolence and pride, shag-eared and deaf, hardly yielding to whip and spur. Now when the charioteer beholds the vision of love, and has his soul warmed through sense, and is full of the prickings and ticklings of desire, the obedient steed, then as always under the government of shame, refrains from leaping on the beloved; but the other, heedless of the pricks and of the blows of the whip, plunges and runs away, giving all manner of trouble to his companion and the charioteer, whom he forces to approach the beloved and to remember the joys of love. They at first indignantly oppose him and will not be urged on to do terrible and unlawful deeds; but at last, when he persists in plaguing them, they yield and agree to do as he bids them. And now they are at the spot and behold the flashing beauty of the beloved; which when the charioteer sees, his memory is carried to the true beauty, whom he beholds in company with Modesty like an image placed upon a holy pedestal. He sees her, but he is afraid and falls backwards in adoration, and by his fall is compelled to pull back the reins with such violence as to bring both the steeds on their haunches, the one willing and unresisting, the unruly one very unwilling; and when they have gone back a little, the one is overcome with shame and wonder, and his whole soul is bathed in perspiration; the other, when the pain is over which the bridle and the fall had given, having with difficulty taken breath, is full of wrath and reproaches, which he heaps upon the charioteer and his fellow steed, for want of courage and manhood, declaring that they have been false to their agreement and guilty of desertion. Again they refuse, and again he urges them on, and will scarce yield to their prayer that he would wait until another time. When the appointed hour comes, they make as if they had forgotten, and he reminds them, fighting and neighing and dragging them on, until at length he, on the same thoughts intent, forces them to draw near again. And when they are near he stoops his head and puts up his tail, and takes the bit in his teeth and pulls shamelessly. Then the charioteer is worse off than ever; he falls back like a racer at the barrier, and with a still more violent wrench drags the bit out of the teeth of the wild steed and covers his abusive tongue and jaws with blood, and forces his legs and haunches to the ground and punishes him sorely. And when this has happened several times and the villain has ceased from his wanton way, he is tamed and humbled, and follows the will of the charioteer, and when he sees the beautiful one he is ready to die of fear. And from that time forward the soul of the lover follows the beloved in modesty and holy fear.

Lucretius

The Roman writer Titus Lucretius Carus (about 99–55 B.C.) is one of the very few who have managed to present philosophical argument in poetic form. His great work *De Rerum Natura* (Of the Nature of Things) gives an eloquent exposition of the materialist philosophy, first sketched by the Greek atomists Leucippus and Democritus (fifth and fourth centuries B.C.), then developed by Epicurus (341–270 B.C.) Plato's arguments for the immaterial nature of the soul are an attempt to refute this ancient materialism. But Lucretius states very clearly some of the classic arguments for the dependence of all mental functioning on the body, for the impossibility of personal survival after death, and for a nonreligious acceptance of our mortality.

These excerpts are from lines 100–900 (approximately) of Book III. The translation from the Latin is by Cyril Bailey (Oxford University Press, 1910).

THE MATERIALITY OF THE SOUL

Now I say that mind and soul are held in union one with the other, and form of themselves a single nature, but that the head, as it were, and lord in the whole body is the reason, which we call mind or understanding, and it is firmly seated in the middle region of the breast. For there it is that fear and terror throb, around these parts are soothing joys; here then is the understanding and the mind. The rest of the soul, spread abroad throughout the body, obeys and is moved at the will and inclination of the understanding. The mind alone by itself has understanding for itself and rejoices for itself, when no single thing stirs either soul or body. And just as, when head or eye hurts within us at the attack of pain, we are not tortured at the same time in all our body; so the mind sometimes feels pain by itself or waxes strong with joy, when all the rest of the soul through the limbs and frame is not roused by any fresh feeling.

60

Nevertheless, when the understanding is stirred by some stronger fear, we see that the whole soul feels with it throughout the limbs, and then sweat and pallor break out over all the body, and the tongue is crippled and the voice is choked, the eyes grow misty, the ears ring, the limbs give way beneath us, and indeed we often see men fall down through the terror in their mind; so that any one may easily learn from this that the soul is linked in union with the mind; for when it is smitten by the force of the mind, straightway it strikes the body and pushes it on.

This same reasoning shows that the nature of mind and soul is bodily. For when it is seen to push on the limbs, to pluck the body from sleep, to change the countenance, and to guide and turn the whole man—none of which things we see can come to pass without touch, nor touch in its turn without body—must we not allow that mind and soul are formed of bodily nature? Moreover, you see that our mind suffers along with the body, and shares its feelings together in the body. If the shuddering shock of a weapon, driven within and laying bare bones and sinews, does not reach the life, yet faintness follows, and a pleasant swooning to the ground, and a turmoil of mind which comes to pass on the ground, and from time to time, as it were, a hesitating will to rise. Therefore it must needs be that the nature of the mind is bodily, since it is distressed by the blow of bodily weapons.

Now of what kind of body this mind is, and of what parts it is formed, I will go on to give account to you in my discourse. First of all I say that it is very fine in texture, and is made and formed of very tiny particles. That this is so, if you give attention, you may be able to learn from this. Nothing is seen to come to pass so swiftly as what the mind pictures to itself coming to pass and starts to do itself. Therefore the mind bestirs itself more quickly than any of the things whose nature is manifest for all to see. But because it is so very nimble, it is bound to be formed of exceeding round and exceeding tiny seeds, so that its particles may be able to move when smitten by a little impulse. For so water moves and oscillates at the slightest impulse, seeing it is formed of little particles, quick to roll. But, on the other hand, the nature of honey is more stable, its fluid more sluggish, and its movement more hesitating; for the whole mass of its matter clings more together, because, we may be sure, it is not formed of bodies so smooth, nor so fine and round. For a light trembling breath can constrain a high heap of poppy-seed to scatter from top to bottom before your eyes: but, on the other hand, a pile of stones or corn-ears it can by no means separate. Therefore, in proportion as bodies are tinier and smoother, so they are gifted with nimbleness. But, on the other hand, all things that are found to be of greater weight or more spiky, the more firm set they are. Now, therefore, since the nature of the mind has been found nimble beyond the rest, it must needs be formed of bodies exceeding small and smooth and round. And this truth, when known to you, will in many things, good friend, prove useful, and will be reckoned of service. This fact, too, declares the nature of the mind, of how thin a texture it is formed, and in how small a place it might be contained, could it be gathered in a mass; that as soon as the unruffled peace of death has laid hold on a man, and the nature of mind and soul has passed away, you could discern nothing there, that sight or weight can test, stolen from the entire body; death preserves all save the feeling of life, and some warm heat. And so it must needs be that the whole soul is made of very tiny seeds, and is linked on throughout veins, flesh, and sinews; inasmuch as, when it is all already gone from the whole body, yet the outer contour of the limbs is preserved unbroken, nor is a jot of weight wanting. Even so it is, when the flavour of wine has passed away or when the sweet breath of a perfume is scattered to the air, or when its savour is gone from some body; still the thing itself seems not a whit smaller to the eyes on that account, nor does anything seem withdrawn from its weight, because, we may be sure, many tiny seeds go to make flavours and scent in the whole body of things. Wherefore once and again you may know that the nature of the understanding and the soul is formed of exceeding tiny seeds, since when it flees away it carries with it no jot of weight. . . .

THE MORTALITY OF THE SOUL

I continue to speak of the soul, proving that it is mortal, suppose that I speak of mind as well, inasmuch as they are at one each with the other and compose a single thing. First of all, since I have shown that it is finely made of tiny bodies and of first-beginnings far smaller than the liquid moisture of water or cloud or smoke— for it far surpasses them in speed of motion, and is more prone to move when smitten by some slender cause; for indeed it is moved by images of smoke and cloud: even as when slumbering in sleep we see altars breathing steam on high, and sending up their smoke; for beyond all doubt these are idols that are borne to us:—now therefore, since, when vessels are shattered, you behold the water flowing away on every side, and the liquid parting this way and that, and since cloud and smoke part asunder into air, you must believe that the soul too is scattered and passes away far more swiftly, and is dissolved more quickly into its first-bodies, when once it is withdrawn from a man's limbs, and has departed. For indeed, since the body, which was, as it were, the vessel of the soul, cannot hold it together, when by some chance it is shattered and made rare, since the blood is withdrawn from the veins, how could you believe that the soul could be held together by any air, which is more rare than our body?

Moreover, we feel that the understanding is begotten along with the body, and grows together with it, and along with it comes to old age. For as children totter with feeble and tender body, so a weak judgement of mind goes with it. Then when their years are ripe and their strength hardened, greater is their sense and increased their force of mind. Afterward, when now the body is shattered by the stern strength of time, and the frame has sunk with its force dulled, then the reason is maimed, the tongue raves, the mind stumbles, all things give way and fail at once. And so it is natural that all the nature of the mind should also be dissolved, even as is smoke, into the high breezes of the air; inasmuch as we see that it is born with the body, grows with it, and, as I have shown, at the same time becomes weary and worn with age.

Then follows this that we see that, just as the body itself suffers wasting diseases and poignant pain, so the mind too has its biting cares and grief and fear; wherefore it is natural that it should also share in death. Nay more, during the diseases of the body the mind often wanders astray; for it loses its reason and speaks raving words, and sometimes in a heavy lethargy is carried off into a deep unending sleep, when eyes and head fall nodding, in which it hears not voices, nor can know the faces of those who stand round, summoning it back to life, bedewing face and cheeks with their tears. Therefore you must needs admit that the mind too is dissolved, inasmuch as the contagion of disease pierces into it. For both pain and disease are alike fashioners of death, as we have been taught ere now by many a man's decease. Again, when the stinging strength of wine has entered into a man, and its heat has spread abroad throughout his veins, why is it that there follows a heaviness in the limbs, his legs are entangled as he staggers, his tongue is sluggish, and his mind heavy, his eyes swim, shouting, sobbing, quarrelling grows apace, and then all the other signs of this sort that go along with them; why does this come to pass, except that the mastering might of the wine is wont to confound the soul even within the body? But whenever things can be so confounded and entangled, they testify that, if a cause a whit stronger shall have made its way within, they must needs perish, robbed of any further life.

When mind and soul then even within the body are tossed by such great maladies, and in wretched plight are rent asunder and distressed, why do you believe that without the body in the open air they can continue life amid the warring winds? And since we perceive that the mind is cured, just like the sick body, and we see that it can be changed by medicine, this too forewarns us that the mind has a mortal life. For whosoever attempts and essays to alter the mind, or seeks to change any other nature, must indeed add parts to it or transfer them from their order, or take away some small whit at least from the whole. But what is immortal does not permit its parts to be transposed, nor that any whit should be added or depart from it. For whenever a thing changes and passes out of its own limits, straightway this is the death of that which was before. And so whether the mind is sick, it gives signs of its mortality as I have proved, or whether it is changed by medicine. So surely is true fact seen to run counter to false reasoning, and to shut off retreat from him who flees, and with double-edged refutation to prove the falsehood. . . .

WE HAVE NOTHING TO FEAR FROM DEATH

Death, then, is naught to us, nor does it concern us a whit, inasmuch as the nature of the mind is but a mortal possession. And even as in the time gone by we felt no ill, when the Poeni came from all sides to the shock of battle, when all the world, shaken by the hurrying turmoil of war, shuddered and reeled beneath the high coasts of heaven, in doubt to which people's sway must fall all human power by land and sea; so, when we shall be no more, when there shall have come the parting of body and soul, by whose union we are made one, you may know that nothing at all will be able to happen to us, who then will be no more, or stir our feeling; no, not if earth shall be mingled with sea, and sea with sky. And even if the nature of mind and the power of soul has feeling, after it has been rent asunder from our body, yet it is naught to us, who are made one by the mating and marriage of body and soul. Nor, if time should gather together our substance after our decease and bring it back again as it is now placed, if once more the light of life should be vouchsafed to us, yet, even were that done, it would not concern us at all, when once the remembrance of our former selves were snapped in twain. And even now we care not at all for the selves that we once were, nor now are we touched by any torturing pain for them. For when you look back over all the lapse of immeasurable time that now is gone, and think how manifold are the motions of matter, you could easily believe this too, that these same seeds, whereof we now are made, have often been placed in the same order as they are now; and yet we cannot recall that in our mind's memory; for in between lies a break in life, and all the motions have wandered everywhere far astray from sense. For, if by chance there is to be grief and pain for a man, he must needs himself too exist at that time, that ill may befall him. Since death forestalls this, and prevents the being of him, on whom these misfortunes might crowd, we may know that we have naught to fear in death, and that he who is no more cannot be wretched, and that it were no whit different if he had never at any time been born, when once immortal death hath stolen away mortal life.

Aristotle

It seems appropriate to place these selections from Aristotle (384–322 B.C.) *after* those from Lucretius, since they express a different understanding of the very concept of mind or soul, which undercuts both the Platonic doctrine that the soul is an incorporeal substance and the materialist view that it is made of atoms. Aristotle's analysis suggests that the soul is not a *thing* at all, whether material or immaterial, but rather a property, or set of properties or abilities, of the living body.

Aristotle's more scientific, indeed biological, approach is seen in his placing of the questions about the soul in the context of living things generally. In his usage, to have a soul means to be alive; and the different kinds of soul are defined by the main categories of life: plants, which merely grow and reproduce; animals, which can move spontaneously and feel sensations; and human beings, who have all these abilities plus the power of thought. Amid the obscure technicalities of Aristotle's text, the best point to focus on is perhaps the suggestion that soul is to body as form is to matter; we can think of the illustrative analogies of the wax and its shape, or the eye and its power of sight.

It would seem to follow from Aristotle's analysis that it is as logically impossible for the powers that constitute having a soul to exist apart from the living body which has them, as it is for the smile of the Cheshire cat to persist after the disappearance of the cat. But Aristotle draws back from inferring this conclusion, at least with respect to the *mind,* or power of thought. He feels there is something special about this power that *could* allow it to exist apart from a living body. He gives no argument here for this exception to what seems to be the clear implication of his general theory. Historically, his hesitation on this point provided an opportunity which Aquinas exploited, as we shall see.

The second selection gives a brief indication of Aristotle's this-worldly, humanist approach to ethics. The final end, that which is worth pursuing for its own sake and is sufficient by itself to make life worth living, is happiness. But what does human happiness consist in? Here Aristotle refers back to his analysis of the different kinds of living thing and concludes that, since rational thought is what distinguishes us from the other animals, our fulfillment lies in the exercise of our rational abilities. In ethics, too, we will find that Aquinas Christianizes Aristotle's concepts.

The reader will notice that Aristotle's thought is more difficult, and his style more dry and technical, than Plato's: he was very much the professional philosopher. But his ideas are so important that no anthology like this can ignore them. The excerpts here are from *de Anima,* Book II (412–413) and *Nichomachean Ethics,* Book I Chapter VII (1097–1098). The translation is from *The Works of Aristotle translated into English,* under the general editorship of W. D. Ross (Oxford University Press, 1937).

HAVING A SOUL IS BEING ALIVE

What has soul in it differs from what has not, in that the former displays life. Now this word has more than one sense, and provided any one alone of these is found in a thing we say that thing is living. Living, that is, may mean thinking or perception or local movement and rest, or movement in the sense of nutrition, decay, and growth. Hence we think of plants also as living, for they are observed to possess in themselves an originative power through which they increase or decrease in all spatial directions; they grow up *and* down, and everything that grows increases its bulk alike in both directions or indeed in all, and continues to live so long as it can absorb nutriment.

This power of self-nutrition can be isolated from the other powers mentioned, but not they from it—in mortal beings at least. The fact is obvious in plants; for it is the only psychic power they possess.

This is the originative power the possession of which leads us to speak of things as *living* at all, but it is the possession of sensation that leads us for the first time to speak of living things as animals; for even those beings which possess no power of local movement but do possess the power of sensation we call animals and not merely living things. . . .

THE INSEPARABILITY OF SOUL FROM BODY

Let us now . . . endeavour to give a precise answer to the question, What is soul? i.e. to formulate the most general possible definition of it.

We are in the habit of recognizing, as one determinate kind of what is, substance, and that in several senses, (a) in the sense of matter or that which in itself is not "a this," and (b) in the sense of form or essence, which is that precisely in virtue of which a thing is called "a this" and thirdly (c) in the sense of that which is compounded of both (a) and (b). Now matter is potentiality, form actuality; of the latter there are two grades related to one another as e.g. knowledge to the exercise of knowledge.

Among substances are by general consent reckoned bodies and especially natural bodies; for they are the principles of all other bodies. Of natural bodies some have life in them, others not; by life we mean self-nutrition and growth (with its correlative decay). It follows that every natural body which has life in it is a substance in the sense of a composite.

But since it is also a *body* of such and such a kind, viz. having life, the *body* cannot be soul; the body is the subject or matter, not what is attributed to it. Hence the soul must be a substance in the sense of the form of a natural body having life potentially within it. But substance is actuality and thus soul is the actuality of a body as above characterized. Now the word actuality has two senses corresponding respectively to the possession of knowledge and the actual exercise of knowledge. It is obvious that the soul is actuality in the first sense, viz. that of knowledge as possessed, for both sleeping and waking presuppose the existence of soul, and of these waking corresponds to actual knowing, sleeping to knowledge possessed but not employed, and, in the history of the individual, knowledge comes before its employment or exercise.

That is why the soul is the first grade of actuality of a natural body having life potentially in it. The body so described is a body which is organized. The parts of plants in spite of their extreme simplicity are "organs"; e.g. the leaf serves to shelter the pericarp, the pericarp to shelter the fruit, while the roots of plants are analogous to the mouth of animals, both serving for the absorption of food. If, then, we have to give a general formula applicable to all kinds of soul, we must describe it as the first grade of actuality of a natural organized body. That is why we can wholly dismiss as unnecessary the question whether the soul and the body are one: it is as meaningless as to ask whether the wax and the shape given to it by the stamp are one, or generally the matter of a thing and that of which it is the matter. . . .

Next, apply this doctrine in the case of the "parts" of the living body. Suppose that the eye were an animal—sight would have been its soul, for sight is the substance or essence of the eye which corresponds to the formula, the eye being merely the matter of seeing; when seeing is removed the eye is no longer an eye, except in name—it is no more a real eye than the eye of a statue or of a painted figure. We must now extend our consideration from the "parts" to the whole living body; for what the departmental sense is to the bodily part which is its organ, that the whole faculty of sense is to the whole sensitive body as such.

We must not understand by that which is "potentially capable of living" what has lost the soul it had, but only what still retains it; but seeds and fruits are bodies which possess the qualification. Consequently, while waking is actuality in a sense corresponding to the cutting and the seeing, the soul is actuality in the sense corresponding to the power of sight and the power in the tool; the body corresponds to what exists in potentiality; as the pupil *plus* the power of sight constitutes the eye, so the soul *plus* the body constitutes the animal.

From this it indubitably follows that the soul is inseparable from its body, or at any rate that certain parts of it are (if it has parts)—for the actuality of some of them is nothing but the actualities of their bodily parts. Yet some may be separable because they are not the actualities of any body at all. . . .

We have no evidence as yet about mind or the power to think; it seems to be a widely different kind of soul, differing as what is eternal from what is perishable; it alone is capable of existence in isolation from all other psychic powers. All the other parts of soul, it is evident from what we have said, are, in spite of certain statements to the contrary, incapable of separate existence though, of course, distinguishable by definition. . . .

THE GOOD FOR MAN

Let us again return to the good we are seeking, and ask what it can be. It seems different in different actions and arts; it is different in medicine, in strategy, and in the other arts likewise. What then is the good of each? Surely that for whose sake everything else is done. In medicine this is health, in strategy victory, in architecture a house, in any other sphere something else, and in every action and pursuit the end; for it is for the sake of this that all men do whatever else they do. Therefore, if there is an end for all that we do, this will be the good achievable by action, and if there are more than one, these will be the goods achievable by action.

So the argument has by a different course reached the same point; but we must try to state this even more clearly. Since there are evidently more than one end, and we choose some of these (e.g. wealth, flutes, and in general instruments) for the sake of something else, clearly not all ends are final ends; but the chief good is evidently something final. Therefore, if there is only one final end, this will be what we are seeking, and if there are more than one, the most final of these will be what we are seeking. Now we call that which is in itself worthy of pursuit more final than that which is worthy of pursuit for the sake of something else, and that which is never desirable for the sake of something else more final than the things that are desirable both in themselves and for the sake of that other thing, and therefore we call final without qualification that which is always desirable in itself and never for the sake of something else.

Now such a thing happiness, above all else, is held to be; for this we choose always for itself and never for the sake of something else, but honour, pleasure, reason, and every virtue we choose indeed for themselves (for if nothing resulted from them we should still choose each of them), but we choose them also for the sake of happiness, judging that by means of them we shall be happy. Happiness, on the other hand, no one chooses for the sake of these, nor, in general, for anything other than itself.

From the point of view of self-sufficiency the same result seems to follow; for the final good is thought to be self-sufficient. Now by self-sufficient we do not mean that which is sufficient for a man by himself, for one who lives a solitary life, but also for parents, children, wife, and in general for his friends and fellow citizens, since man is born for citizenship. But some limit must be set to this; for if we extend our requirement to ancestors and descendants and friends' friends we are in for an infinite series. Let us examine this question, however, on another occasion; the self-sufficient we now define as that which when isolated makes life desirable and lacking in nothing; and such we think happiness to be; and further we think it most desirable of all things, without being counted as one good thing among others—if it were so counted it would clearly be made more desirable by the addition of even the least of goods; for that which is added becomes an excess of goods, and of goods the greater is always more desirable. Happiness, then, is something final and self-sufficient, and is the end of action.

Presumably, however, to say that happiness is the chief good seems a platitude, and a clearer account of what it is is still desired. This might perhaps be given, if we could first ascertain the function of man. For just as for a flute-player, a sculptor, or any artist, and, in general, for all things that have a function or activity, the good and the "well" is thought to reside in the function, so would it seem to be for man, if he has a function. Have the carpenter, then, and the tanner certain functions or activities, and has man none? Is he born without a function? Or as eye, hand, foot, and in general each of the parts evidently has a function, may one lay it down that man similarly has a function apart from all these? What then can this be? Life seems to be common even to plants, but we are seeking what is peculiar to man. Let us exclude, therefore, the life of nutrition and growth. Next there would be a life of perception, but *it* also seems to be common even to the horse, the ox, and every animal. There remains, then, an active life of the element that has a rational principle; of this, one part has such a principle in the sense of being obedient to one, the other in the sense of possessing one and exercising thought. And, as "life of the rational element" also has two meanings, we must state that life in the sense of activity is what we mean; for this seems to be the more proper sense of the term. Now if the function of man is an activity of soul which follows or implies a rational principle, and if we say "a so-and-so" and "a good so-and-so" have a function which is the same in kind, e.g. a lyre-player and a good lyre-player, and so without qualification in all cases, eminence in respect of goodness being added to the name of the function (for the function of a lyre-player is to play the lyre and that of a good lyre-player is to do so well): if this is the case, [and we state the function of man to be a certain kind of life, and this to be an activity or actions of the soul implying a rational principle, and the function of a good man to be the good and noble performance of these, and if any action is well performed when it is performed in accordance with the appropriate excellence: if this is the case,] human good turns out to be activity of soul in accordance with virtue, and if there are more than one virtue, in accordance with the best and most complete.

But we must add "in a complete life." For one swallow does not make a summer, nor does one day; and so too one day, or a short time, does not make a man blessed and happy.

Aquinas

In the first few centuries after Christ, there took place a creative fusion of ideas from the Greek philosophers, especially Plato and the mystical neo-Platonists such as Plotinus, with the religious beliefs of the Christian writers of the New Testament. Thus, the Platonic dualism of body and soul, with its ethical implication that care of the soul is more important than anything to do with the body, found a ready home in the thought of the Church Fathers, such as St. Augustine.

The works of Aristotle were almost completely lost to the Christian West until they were rediscovered in the late Middle Ages, via the Arabs. They then exerted an enormous influence, so much so that we find St. Thomas Aquinas (about 1224–1274) referring to Aristotle simply as "the Philosopher." In his massive *Summa Theologica* and other works, Aquinas built an impressive intellectual synthesis of Aristotelean and Christian ideas, which was new and controversial at the time, but has since become Roman Catholic orthodoxy.

Aquinas makes use of Aristotle's reservations about the power of thought to attempt a curious compromise between Platonic, Aristotelean, and Biblical theories of human nature. He follows Aristotle's main line in insisting that the soul is not a body, but the act or property of a body. He then offers an argument for the special treatment of what he calls "the principle of intellectual operation," which he views as something that is "self-subsistent" and could, like Plato's soul, exist apart from the body. Having argued that a future life is necessary if true and perfect happiness is to be possible, Aquinas asserts the doctrine of the resurrection of individual people (which we have seen in the New Testament passage from St. Paul). Then, he is able to use the Platonic element in his philosophy to answer the question of what makes the resurrected body the *same person* as had died, asserting that the resurrection consists in joining the same separated soul to the same body. Whether this eclectic theory makes sense, readers must judge.

In the *Summa Theologica,* Aquinas has a complicated mode of proceeding—in each section, he poses a question, states objections to the view he considers correct, quotes authority from the Bible, the Church Fathers, or Aristotle for his own position, then gives philosophical arguments for it, and finally answers the original objections. In these selections, the arguments are excerpted from Part I, Q. LXXV art. 1, art. 2, Q. LXXXIII art. 1; Part II (1st section) Q. I art. 1, art. 7, art. 8, Q. V art. 3; Supplement Q. LXXXIX art. 2. The translation from the Latin is by the Fathers of the English Dominican Province (R. & T. Washbourne Ltd., 1912).

THE SOUL IS AN ACT OF THE BODY

To seek the nature of the soul, we must premise that the soul is defined as the first principle of life in those things which live: for we call living things animate (souled): and those things which have no life, inanimate (soulless). Now life is shown principally by two actions, knowledge and movement. The philosophers of old, not being able to rise above their imagination, supposed that the principle of these actions was something corporeal; for they asserted that only bodies were real things; and that what is not corporeal is nothing: hence they maintained that the soul is something corporeal. This opinion can be proved to be false in many ways; but we shall only make use of one proof, based on universal and certain principles, which shows clearly that the soul is not a body.

It is manifest that not every principle of vital action is a soul, for then the eye would be a soul, as it is a principle of vision; and the same might be applied to the other instruments of the soul; but it is the *first* principle of life, which we call the soul. Now, though a body may be a principle of life, as the heart is a principle of life in an animal; yet nothing corporeal can be the first principle of life. For it is clear that to be a principle of life, or to be a living thing, does not belong to a body as such; since, if that were the case, everything corporeal would be a living thing, or a principle of life. Therefore a body is a living thing or even a principle of life, as *such* a body. Now that it is actually such a body, it owes to some principle which is called its act. Therefore the soul, which is the first principle of life, is not a body, but the act of a body; as heat, which is the principle of calefaction, is not a body, but an act of a body. . . .

. . . BUT CAN EXIST SEPARATELY

It must necessarily be allowed that the principle of intellectual operation, which we call the soul, is a principle both incorporeal and subsistent. For it is clear that by means of the intellect man can have knowledge of all corporeal things. Now whatever knows certain things cannot have any of them in its own nature; because that which is in it naturally would impede the knowledge of other things. Thus we observe that a sick man's tongue being vitiated by a feverish and bitter humour, is insensible to anything sweet, and everything seems bitter to it. Therefore, if the intellectual principle had something corporeal in its nature, it would be unable to know all bodies. Now every body has its own determinate nature. Therefore it is impossible for the intellectual principle to be a body. It is likewise impossible for it to understand by means of a bodily organ; since the determinate nature of that organ would impede knowledge of all bodies; as when a certain determinate colour is not only in the pupil of the eye, but also in a glass vase, the liquid in the vase seems to be of that same colour. Therefore the intellectual principle which we call the mind or the intellect has an operation of its own apart from the body. Now only a self-subsisting thing can have an operation of its own; for nothing can operate but what is actual. . . .

THE FREEDOM OF THE WILL

Man has free-will: otherwise counsels, exhortations, commands, prohibitions, rewards and punishments would be in vain. In order to make this evident, we must observe that some things act without judgment; as a stone moves downwards; and in like manner all things which lack knowledge. And some act from judgment, but not a free judgment; as brute animals. For the sheep, seeing the wolf, judges it a thing to be shunned, from a natural and not a free judgment, because it judges, not from reason, but from natural instinct. And the same thing is to be said of any judgment of brute animals. But man acts from judgment, because by his apprehensive power he judges that something should be avoided or sought. But because this judgment, in the case of some particular act, is not from a natural instinct, but from some act of comparison in the reason, therefore he acts from free judgment and retains the power of being inclined to various things. For reason in contingent matters may follow opposite courses, as we see in dialectic, syllogisms and rhetorical arguments. Now particular operations are contingent, and therefore in such matters the judgment of reason may follow opposite courses, and is not determinate to one. And forasmuch is it necessary that that man has a free-will, as he is rational. . . .

Of actions done by man those alone are properly called *human*, which are proper to man as man. Now man differs from irrational animals in this, that he is master of his actions. Wherefore those actions alone are properly called human, of which man is master. Now man is master of his actions through his reason and will; whence too, the free-will is called the faculty of will and reason. Therefore those actions are properly called human, which proceed from a deliberate will. And if any other actions befit man, they can be called actions *of a man,* but not properly *human* actions, since they are not proper to man as man.—Now it is clear that whatever actions proceed from a power, are caused by that power in accordance with the nature of its object. But the object of the will is the end and the good. Therefore all human actions must be for an end. . . .

THE PURPOSE OF HUMAN LIFE

Augustine says (*De Trin,* xiii.) that all men agree in desiring the last end, which is happiness.

We can speak of the last end in two ways: first, considering only the aspect of last end; secondly, considering the thing in which the aspect of last end is realized. So, then, as to the aspect of last end, all agree in desiring the last end; since all desire the fulfilment of their perfection, and it is precisely this fulfilment in which the last end consists, as stated above. But as to the thing in which this aspect is realized, all men are not agreed as to their last end: since some desire riches, as their consummate good; some, pleasure; others, something else. Thus to every taste the sweet is pleasant; but to some, the sweetness of wine is most pleasant, to others, the sweetness of honey, or of something similar. Yet that sweet is absolutely the best of all pleasant things, in which he who has the best taste takes most pleasure. In like manner that good is most complete which the man with well-disposed affections desires for his last end. . . .

[M]an's last end is happiness; which all men desire, as Augustine says (*De Trin.* xiii.). But *happiness is not possible for animals bereft of reason*, as Augustine says.

As the Philosopher says (*Phys.* ii. and *Metaph.* v.), the end is twofold,—the end *for which* and the end *by which;* viz., the thing itself in which is found the aspect of good, and the use or acquisition of that thing. Thus we say that the end of the movement of a weighty body is either a lower place as *thing,* or to be in a lower place, as *use;* and the end of the miser is money as *thing,* or possession of money as *use.*

If, therefore, we speak of man's last end as of the thing which is the end, thus all other things concur in man's last end, since God is the last end of man and of all other things.—If, however, we speak of man's last end, as of the acquisition of the end, then irrational creatures do not concur with man in this end. For man and other rational creatures attain to their last end by knowing and loving God: this is not possible to other creatures, which acquire their last end, in so far as they share in the Divine likeness, inasmuch as they are, or live, or even know. . . .

A certain participation of Happiness can be had in this life: but perfect and true Happiness cannot be had in this life. This may be seen from a twofold consideration.

First, from the general notion of happiness. For since happiness is a *perfect and sufficient good,* it excludes every evil, and fulfils every desire. But in this life every evil cannot be excluded. For this present life is subject to many unavoidable evils; to ignorance on the

part of the intellect; to inordinate affection on the part of the appetite, and to many penalties on the part of the body; as Augustine
sets forth in *De Civ. Dei.* chap. iv. Likewise neither can the desire
for good be satiated in this life. For man naturally desires the good,
which he has, to be abiding. Now the goods of the present life pass
away; since life itself passes away, which we naturally desire to
have, and would wish to hold abidingly, for man naturally shrinks
from death. Wherefore it is impossible to have true Happiness in
this life.

Secondly, from a consideration of the specific nature of Happiness, viz., the vision of the Divine Essence, which man cannot
obtain in this life, as was shown in the First Part. Hence it is evident
that none can attain true and perfect Happiness in this life. . . .

THE RESURRECTION OF THE BODY

It is written (Job. xix. 27): *Whom I myself shall see . . . and not
another,* and he is speaking of the vision after the resurrection.
Therefore the same identical man will rise again.

Further, Augustine says (*De Trin.* viii. 5) that *to rise again is
naught else but to live again.* Now unless the same identical man that
died return to life, he would not be said to live again. Therefore he
would not rise again, which is contrary to faith.

The necessity of holding the resurrection arises from this, —that
man may obtain the last end for which he was made; for this cannot
be accomplished in this life, nor in the life of the separated soul, as
stated above: otherwise man would have been made in vain, if he
were unable to obtain the end for which he was made. And since it
behooves the end to be obtained by the selfsame thing that was
made for that end, lest it appear to be made without purpose, it is
necessary for the selfsame man to rise again; and this is effected by
the selfsame soul being united to the selfsame body. For otherwise
there would be no resurrection properly speaking, if the same man
were not reformed. Hence to maintain that he who rises again is
not the selfsame man is heretical, since it is contrary to the truth of
Scripture which proclaims the resurrection.

Part III

The Search
for a Scientific Theory of
Human Nature

Since the Industrial Revolution of the nineteenth century, science has transformed the conditions of life for those fortunate enough to share its benefits. The modern, practical applications of science are based on the earlier intellectual achievement of scientific explanations of the physical world. The roots of science can be traced back to the Greeks, but systematic *observation* and *experimentation,* with the aim of testing old theories and perhaps developing new ones, did not emerge until the period when the medieval synthesis began to break down and give way to the ferment of new ideas in the Renaissance. The so-called "scientific revolution" is generally agreed to have reached its culmination in the seventeenth century. Newton's theory of mechanics, which successfully explained the movements of all material things (on earth *and* in the heavens) by a few simple mathematical laws, became a paradigm for all science to follow.

From the seventeenth century onwards, science acquired so much prestige that it became a part of the intellectual outlook of every educated person. Even before Newton's grand synthesis in the 1660s, thinkers were speculating about the application of the new scientific methods to human nature and society. In the 1620s, William Harvey had demonstrated the circulation of the blood through the body, thus showing how there was substantial new knowledge to be gained about ourselves by the methods of science. The question of how a complete physical explanation of the workings of our bodies could be reconciled with our view of ourselves as free agents, and as having distinctively mental powers of rational thought, immediately became pressing. Indeed, it is no exaggeration to say that the main theme of philosophy since the seventeenth century has been this problem of the relationship between scientific and other understandings of human nature.

SEARCH FOR A SCIENTIFIC THEORY OF HUMAN NATURE

We shall now look at a whole series of attempts to develop a scientific theory of human nature. Some of them are more explicitly methodological and philosophical, trying to show what form such a theory should take, and arguing about what it would and would not imply. Others do more in the way of developing a substantive theory, although, as we shall see, they find it impossible to avoid methodological and philosophical problems, which dog the footsteps of the would-be scientists of human nature right down to the present day.

René Descartes

The Frenchman René Descartes (1596–1650) was a central figure in the scientific revolution, being a mathematician, experimental scientist, and philosopher. In mathematics, he invented the Cartesian coordinates which allowed geometry to be expressed in terms of algebra; and in mechanics, cosmology, and physiology, his theories and research helped to push forward those sciences. His philosophical dualism between body and soul provided an obvious solution to the problems involved in applying science to man, because the body could be understood as the subject of a deterministic, mechanical explanation, while the distinctively human attributes of thought, rationality, and freedom could be located in the incorporeal soul, beyond all reach of science.

This dualist metaphysics is clearly in agreement with Plato and with the Roman Catholic orthodoxy to which Descartes subscribed. But he gave new arguments for the old position: his famous "cogito, ergo sum" takes as its premise his indubitable awareness of his own thinking; he contrasts this with the alleged doubtfulness of all claims about material objects, even his own body, and concludes that he (i.e., his soul) must be an incorporeal substance. This supposedly incorrigible awareness of the contents of our own consciousness, on which Descartes laid so much stress, was widely assumed to rule out any form of materialist theory of mind, until questioned in the twentieth century by behaviorists such as J. B. Watson, and identity theorists such as H. Feigl (both discussed below).

Descartes' dualism forced him to make an absolute distinction between those who possessed incorporeal souls and that which lacked them, and he drew this line between men and all other animals. Having tied all mental attributes to the soul, he thus had to treat the beasts as mere bodily machines, lacking not only rationality but even sensation (unlike Aristotle). As shown in the extract here, he thought that this absolute distinction was supported by two tests which show the presence of "rational souls" in men alone—our use of language, and the all-round nature of our intelligence. Later, we shall see Descartes' appeal to the uniqueness of human language taken up by Chomsky.

These excerpts are from Parts IV and V of the *Discourse on Method* of 1637, in which Descartes gave a preliminary summary of his research. The translation from the French is by Anscombe and Geach (in Descartes, *Philosophical Writings,* Nelson, London; in the United States, Bobbs-Merrill, 1971).

CONSCIOUSNESS, AND DUALISM

I do not know whether I need tell you of my first meditations; for they are perhaps too metaphysical and uncommon for the general taste. At the same time I am in a way obliged to speak of them so as to make it possible to judge whether the foundation I have chosen is secure enough. I had noticed long before, as I said just now, that in conduct one sometimes has to follow opinions that one knows to be most uncertain just as if they were indubitable; but since my present aim was to give myself up to the pursuit of truth alone, I thought I must do the very opposite, and reject as if absolutely false anything as to which I could imagine the least doubt, in order to see if I should not be left at the end believing something that was absolutely indubitable. So, because our senses sometimes deceive us, I chose to suppose that nothing was such as they lead us to imagine. Because there are men who make mistakes in reasoning even as regards the simplest points of geometry and perpetrate fallacies, and seeing that I was as liable to error as anyone else, I rejected as false all the arguments I had so far taken for demonstrations. Finally, considering that the very same experiences (*pensées*) as we have in waking life may occur also while we sleep, without there being at that time any truth in them, I decided to feign that everything that had entered my mind hitherto was no more true than the illusions of dreams. But immediately upon this I noticed that while I was trying to think everything false, it must needs be that I, who was thinking this (*qui le pensais*), was something. And observing that this truth "I am thinking (*je pense*), therefore I exist" was so solid and secure that the most extravagant suppositions of the sceptics could not overthrow it, I judged that I need not scruple to accept it as the first principle of philosophy that I was seeking.

I then considered attentively what I was; and I saw that while I could feign that I had no body, that there was no world, and no place existed for me to be in, I could not feign that I was not; on the contrary, from the mere fact that I thought of doubting (*je pensais à douter*) about other truths it evidently and certainly followed that I existed. On the other hand, if I had merely ceased to be conscious (*de penser*), even if everything else that I had ever imagined had been true, I had no reason to believe that I should still have existed. From this I recognised that I was a substance whose whole essence or nature is to be conscious (*de penser*) and whose being requires no place and depends on no material thing. Thus this self (*moi*), that is to say the soul, by which I am what I am, is entirely distinct from the body, and is even more easily known; and even if the body were not there at all, the soul would be just what it is. . . .

LANGUAGE AND RATIONALITY

I went on to describe animals and in particular men. But since I had not yet enough knowledge to speak of them in the same style as other objects, namely by demonstrating effects from their causes and showing from what seeds and in what manner Nature must produce them, I confined myself to imagining that God should form a human body just like our own both in the outward shape of its limbs and in the interior arrangement of its organs, without using any matter but what I had described, and without placing in it, to begin with, any rational soul, or anything to serve as a vegetative or sensitive soul. . . . Examining the functions that might result in such a body, what I found were precisely those that may occur in us unconsciously, without any co-operation of the soul, that is to say of the element distinct from the body of which I said above that its nature is merely to be conscious; the very operations in which irrational animals resemble us; but I could find none of the operations that depend on consciousness and are alone proper to us as men; whereas I could find a place for these on the further supposition that God created a rational soul, and joined it to the body in a way that I described. . . .

I specially dwelt on showing that if there were machines with the organs and appearance of a monkey, or some other irrational animal, we should have no means of telling that they were not altogether of the same nature as those animals; whereas if there were machines resembling our bodies, and imitating our actions as far as is morally possible, we should still have two means of telling that, all the same, they were not real men. First, they could never use words or other constructed signs, as we do to declare our thoughts to others. It is quite conceivable that a machine should be so made as to utter words, and even utter them in connexion with physical events that cause a change in one of its organs; so that e.g. if it is touched in one part, it asks what you want to say to it, and if touched in another, it cries out that it is hurt; but not that it should be so made as to arrange words variously in response to the meaning of what is said in its presence, as even the dullest men can do. Secondly, while they might do many things as well as any of us or better, they would infallibly fail in others, revealing that they acted not from knowledge but only from the disposition of their organs. For while reason is a universal tool that may serve in all kinds of circumstances, these organs need a special arrangement for each special action; so it is morally impossible that a machine should contain so many varied arrangements as to act in all the events of life in the way reason enables us to act.

Now in just these two ways we can also recognise the difference between men and brutes. For it is a very remarkable thing that there are no men so dull and stupid, not even lunatics, that they cannot arrange various words and form a sentence to make their thoughts (*pensées*) understood; but no other animal, however perfect or well bred, can do the like. This does not come from their lacking the organs; for magpies and parrots can utter words like ourselves, and yet they cannot talk like us, that is, with any sign of being aware of (*qu'ils pensent*) what they say. Whereas men born deaf-mutes, and thus devoid of the organs that others use for speech, as much as brutes are or more so, usually invent for themselves signs by which they make themselves understood to those who are normally with them, and who thus have a chance to learn their language. This is evidence that brutes not only have a smaller degree of reason than men, but are wholly lacking in it. For it may be seen that a very small degree of reason is needed in order to be able to talk; and in view of the inequality that occurs among animals of the same species, as among men, and of the fact that some are easier to train than others, it is incredible that a monkey or parrot who was one of the most perfect members of his species should not be comparable in this regard to one of the stupidest children or at least to a child with a diseased brain, if their souls were not wholly different in nature from ours. And we must not confuse words with natural movements, the expressions of emotion, which can be imitated by machines as well as by animals. Nor must we think, like some of the ancients, that brutes talk but we cannot understand their language; for if that were true, since many of their organs are analogous to ours, they could make themselves understood to us, as well as to their fellows. It is another very remarkable thing that although several brutes exhibit more skill than we in some of their actions, they show none at all in many other circumstances; so their excelling us is no proof that they have a mind (*de l'esprit*), for in that case they would have a better one than any of us and would excel us all round; it rather shows that they have none, and that it is nature that acts in them according to the arrangements of their organs; just as we see how a clock, composed merely of wheels and springs, can reckon the hours and measure time more correctly than we can with all our wisdom.

I went on to describe the rational soul, and showed that, unlike the other things I had spoken of, it cannot be extracted from the potentiality of matter, but must be specially created; and how it is not enough for it to dwell in the human body like a pilot in his ship, which would only account for its moving the limbs of the body; in order to have in addition feelings and appetites like ours, and so make up a true man, it must be joined and united to the body more closely. Here I dwelt a little on the subject of the soul, as among the most important; for, after the error of denying God, (of which I think I have already given a sufficient refutation), there is none more likely to turn weak characters from the strait way of virtue than the supposition that the soul of brutes must be of the same nature as ours, so that after this life we have no more to hope or fear than flies or ants. Whereas, when we realise how much they really differ from us, we understand much better the arguments proving that our soul is of a nature entirely independent of the body, and thus not liable to die with it; and since we can discern no other causes that should destroy it, we are naturally led to decide that it is immortal.

Thomas Hobbes

The Englishman Thomas Hobbes (1588–1679) published his *Leviathan* in 1651, just after the English civil war. It is famous mainly as political philosophy, arguing the need for a sovereign authority which wields supreme power, to save men from the evils of the "state of nature" in which every individual is at risk from the rest. But, as we shall see here, in these excerpts from the Introduction, Part I, Chapters 6 and 13, and Part II, Chapter 17, this political conclusion is derived from premises about individual human nature. Hobbes' materialism is exhibited in his treatment of life as a motion of the limbs, of sensation as a motion in the organs within the body, and of desire as the inner cause of bodily movement. There is no mention of soul (indeed, he elsewhere declared the notion of incorporeal substance to be a contradiction in terms). Whatever lip service Hobbes may have felt constrained to pay to the religious views of his time, we have here a sketch of an uncompromisingly materialist theory of human nature.

MATERIALISM

Nature, the art whereby God hath made and governs the world, is by the *art* of man, as in many other things, so in this also imitated, that it can make an artificial animal. For seeing life is but a motion of limbs, the beginning whereof is in some principal part within; why may we not say, that all *automata* (engines that move themselves by springs and wheels as doth a watch) have an artificial life? For what is the *heart,* but a *spring;* and the *nerves,* but so many *strings;* and the *joints,* but so many *wheels,* giving motion to the whole body, such as was intended by the artificer? *Art* goes yet further, imitating that rational and most excellent work of nature, *man.* For by art is created that great Leviathan called a Commonwealth, or State, in Latin Civitas, which is but an artificial man; though of greater stature and strength than the natural, for whose protection and defence it was intended; and in which the *sovereignty* is an artificial *soul,* as giving life and motion to the whole body; the *magistrates,* and other *officers* of judicature and execution, artificial *joints; reward and punishment,* by which fastened to the seat of the sovereignty every joint and member is moved to perform his duty, are the *nerves,* that do the same in the body natural; the *wealth* and *riches* of all the particular members, are the *strength; salus populi,* the *people's safety,* its *business; counsellors,* by whom all things needful for it to know are suggested unto it, are the *memory; equity,* and *laws,* an artificial *reason* and *will; concord, health; sedition, sickness;* and *civil war, death.* Lastly, the *pacts* and *covenants,* by which the parts of this body politic were at first made, set together, and united, resemble that *fiat,* or the *let us make man,* pronounced by God in the creation.

To describe the nature of this artificial man, I will consider:

First, the *matter* thereof, and the *artificer;* both which is *man.*

Secondly, *how,* and by what *covenants* it is made; what are the *rights* and just *power* or *authority* of a *sovereign;* and what it is that *preserveth* and *dissolveth it.* . . .

There be in animals, two sorts of *motions* peculiar to them: one called *vital;* begun in generation, and continued without interruption through their whole life; such as are the *course* of the *blood,* the *pulse,* the *breathing,* the *concoction, nutrition, excretion,* etc., to which motions there needs no help of imagination: the other is *animal motion,* otherwise called *voluntary motion;* as to *go,* to *speak,* to *move* any of our limbs, in such manner as is first fancied in our minds. That sense is motion in the organs and interior parts of man's body, caused by the action of the things we see, hear, etc.; and that fancy is but the relics of the same motion, remaining after sense, has been already said in the first and second chapters. And because *going, speaking,* and the like voluntary motions, depend always upon a precedent thought of *whither, which way,* and *what;* it is evident, that the imagination is the first internal beginning of all voluntary motion. And although unstudied men do not conceive any motion at all to be there where the thing moved is invisible; or the space it is moved in is, for the shortness of it, insensible; yet that doth not hinder, but that such motions are. For let a space be never so little, that which is moved over a greater space, whereof that little one is part, must first be moved over that. These small beginnings of motion, within the body of man, before they appear in walking, speaking, striking, and other visible actions, are commonly called ENDEAVOUR.

This endeavour, when it is toward something which causes it, is called APPETITE, or DESIRE; the latter, being the general name; and the other oftentimes restrained to signify the desire of food, namely *hunger* and *thirst*. And when the endeavour is fromward something, it is generally called AVERSION. . . .

Of appetites and aversions, some are born with men; as appetite of food, appetite of excretion, and exoneration, which may also and more properly be called aversions, from somewhat they feel in their bodies; and some other appetites, not many. The rest, which are appetites of particular things, proceed from experience, and trial of their effects upon themselves or other men. For of things we know not at all, or believe not to be, we can have no further desire, than to taste and try. But aversion we have for things, not only which we know have hurt us, but also that we do not know whether they will hurt us, or not. . . .

But whatsoever is the object of any man's appetite or desire, that is it which he for his part calleth *good:* and the object of his hate and aversion, *evil;* and of his contempt, *vile* and *inconsiderable.* For these words of good, evil, and contemptible, are ever used with relation to the person that useth them: there being nothing simply and absolutely so; nor any common rule of good and evil, to be taken from the nature of the objects themselves; but from the person of the man, where there is no commonwealth; or, in a commonwealth, from the person that representeth it; or from an arbitrator or judge, whom men disagreeing shall by consent set up, and make his sentence the rule thereof. . . .

CONFLICT BETWEEN MEN

Nature hath made men so equal, in the faculties of the body, and mind; as that though there be found one man sometimes manifestly stronger in body, or of quicker mind than another; yet when all is reckoned together, the difference between man, and man, is not so considerable, as that one man can thereupon claim to himself any benefit, to which another may not pretend, as well as he. For as to the strength of body, the weakest has strength enough to kill the strongest, either by secret machination, or by confederacy with others, that are in the same danger with himself.

And as to the faculties of the mind, setting aside the arts grounded upon words, and especially that skill of proceeding upon general, and infallible rules, called science; which very few have, and but in few things; as being not a native faculty, born with us; nor attained, as prudence, while we look after somewhat else, I find yet a greater equality amongst men, than that of strength. For prudence is but experience; which equal time, equally bestows on all men, in those things they equally apply themselves unto. That which may perhaps make such equality incredible, is but a vain conceit of one's own wisdom, which almost all men think they have in a greater degree, than the vulgar; that is, than all men but themselves, and a few others, whom by fame, or for concurring with themselves, they approve. For such is the nature of men, that howsoever they may acknowledge many others to be more witty, or more eloquent, or more learned; yet they will hardly believe there be many so wise as themselves; for they see their own wit at hand, and other men's at a distance. But this proveth rather that men are in that point equal, than unequal. For here is not ordinarily a greater sign of the equal distribution of any thing, than that every man is contented with his share.

From this equality of ability, ariseth equality of hope in the attaining of our ends. And therefore if any two men desire the same thing, which nevertheless they cannot both enjoy, they become enemies; and in the way to their end, which is principally their own conservation, and sometimes their delectation only, endeavour to destroy, or subdue one another. And from hence it comes to pass, that where an invader hath no more to fear, than another man's single power; if one plant, sow, build, or possess a convenient seat, others may probably be expected to come prepared with forces united, to dispossess, and deprive him, not only of the fruit of his labour, but also of his life, or liberty. And the invader again is in the like danger of another.

And from this diffidence of one another, there is no way for any man to secure himself, so reasonable, as anticipation; that is, by force, or wiles, to master the persons of all men he can, so long, till he see no other power great enough to endanger him: and this is no more than his own conservation requireth, and is generally allowed. Also because there be some, that taking pleasure in contemplating their own power in the acts of conquest, which they pursue farther than their security requires; if others, that otherwise would be glad to be at ease within modest bounds, should not by invasion increase their power, they would not be able, long time, by standing only on their defence, to subsist. And by consequence, such augmentation of dominion over men being necessary to a man's conservation, it ought to be allowed him.

Again, men have no pleasure, but on the contrary a great deal of grief, in keeping company, where there is no power able to over-awe them all. For every man looketh that his companion should value him, at the same rate he sets upon himself: and upon all signs of contempt, or undervaluing, naturally endeavours, as far as he dares, (which amongst them that have no common power to keep them in quiet, is far enough to make them destroy each other), to extort a greater value from his contemners, by damage; and from others, by the example.

So that in the nature of man, we find three principal causes of quarrel. First, competition; secondly, diffidence; thirdly, glory.

The first, maketh men invade for gain; the second, for safety; and the third, for reputation. The first use violence, to make themselves masters of other men's persons, wives, children, and cattle; the second, to defend them; the third, for trifles, as a word, a smile, a different opinion, and any other sign of undervalue, either direct in their persons, or by reflection in their kindred, their friends, their nation, their profession, or their name.

Hereby it is manifest, that during the time men live without a common power to keep them all in awe, they are in that condition which is called war; and such a war, as is of every man, against every man. For WAR, consisteth not in battle only, or the art of fighting; but in a tract of time, wherein the will to contend by battle is sufficiently known: and therefore the notion of *time,* is to be considered in the nature of war; as it is in the nature of weather. For as the nature of foul weather, lieth not in a shower or two of rain; but in an inclination thereto of many days together: so the nature of war, consisteth not in actual fighting; but in the known disposition thereto, during all the time there is no assurance to the contrary. All other time is PEACE.

Whatsoever therefore is consequent to a time of war, where every man is enemy to every man; the same is consequent to the time, wherein men live without other security, than what their own strength, and their own invention shall furnish them withal. In such condition, there is no place for industry; because the fruit thereof is uncertain: and consequently no culture of the earth; no navigation, nor use of the commodities that may be imported by sea; no commodious building; no instruments of moving, and removing, such things as require much force; no knowledge of the face of the earth; no account of time; no arts; no letters; no society; and which is worst of all, continual fear, and danger of violent death; and the life of man, solitary, poor, nasty, brutish, and short. . . .

THE NEED FOR GOVERNMENT

It is true, that certain living creatures, as bees, and ants, live sociably one with another, which are therefore by Aristotle numbered amongst political creatures; and yet have no other direction, than their particular judgments and appetites; nor speech, whereby one of them can signify to another, what he thinks expedient for the common benefit: and therefore some man may perhaps desire to know, why mankind cannot do the same. To which I answer,

First, that men are continually in competition for honour and dignity, which these creatures are not; and consequently amongst men there ariseth on that ground, envy and hatred, and finally war; but amongst these not so.

Secondly, that amongst these creatures, the common good differeth not from the private; and being by nature inclined to their private, they procure thereby the common benefit. But man, whose joy consisteth in comparing himself with other men, can relish nothing but what is eminent.

Thirdly, that these creatures, having not, as man, the use of reason, do not see, nor think they see any fault, in the administration of their common business: whereas amongst men, there are very many, that think themselves wiser, and abler to govern the public, better than the rest; and these strive to reform and innovate, one this way, another that way; and thereby bring it into distraction and civil war.

Fourthly, that these creatures, though they have some use of voice, in making known to one another their desires, and other affections; yet they want that art of words, by which some some men can represent to others, that which is good, in the likeness of evil; and evil, in the likeness of good; and augment, or diminish the apparent greatness of good and evil; discontenting men, and troubling their peace at their pleasure.

Fifthly, irrational creatures cannot distinguish between *injury,* and *damage;* and therefore as long as they be at ease, they are not offended with their fellows: whereas man is then most troublesome, when he is most at ease: for then it is that he loves to shew his wisdom, and control the actions of them that govern the commonwealth.

Lastly, the agreement of these creatures is natural; that of men, is by covenant only, which is artificial: and therefore it is no wonder if there be somewhat else required, besides covenant, to make their agreement constant and lasting; which is a common power, to keep them in awe, and to direct their actions to the common benefit.

The only way to erect such a common power, as may be able to defend them from the invasion of foreigners, and the injuries of one another, and thereby to secure them in such sort, as that by their own industry, and by the fruits of the earth, they may nourish themselves and live contentedly; is, to confer all their power and strength upon one man, or upon one assembly of men, that may reduce all their wills, by plurality of voices, unto one will: which is as much as to say, to appoint one man, or assembly of men, to bear their person; and every one to own, and acknowledge himself to be author of whatsoever he that so beareth their person, shall act, or cause to be acted, in those things which concern the common peace and safety; and therein to submit their wills, every one to his will, and their judgments, to his judgment. This is more than consent, or concord; it is a real unity of them all, in one and the same person made by covenant of every man with every man, in such manner, as if every man should say to every man, *I authorize and give up my right of governing myself, to this man, or to this assembly of men, on this condition, that thou give up thy right to him, and authorize all his actions in like manner.* This done, the multitude so united in one person, is called a COMMONWEALTH, in Latin CIVITAS. This is the generation of that great LEVIATHAN, or rather, to speak more reverently, of that *mortal god,* to which we owe under the *immortal God,* our peace and defence. For by this authority, given him by every particular man in the commonwealth, he hath the use of so much power and strength conferred on him, that by terror thereof, he is enabled to form the wills of them all, to peace at home, and mutual aid against their enemies abroad. And in him consisteth the essence of the commonwealth; which, to define it, is *one person, of whose acts a great multitude, by mutual covenants one with another, have made themselves every one the author, to the end he may use the strength and means of them all, as he shall think expedient, for their peace and common defence.*

Baruch Spinoza

As Aristotle is to Plato and Lucretius, proposing an alternative to dualism and crude materialism, so Spinoza is to Descartes and Hobbes. But, also like Aristotle, the thought and style of expression in Spinoza are very difficult, and it is hard to get beyond merely repeating his slogan that mind and body are the same thing, conceived under the different attributes of thought and extension (i.e., occupancy of space). Spinoza (1632–1677) was a Dutch Jew, who was excommunicated for unorthodoxy: he was notorious for his identification of God with the whole of Nature. In his *Ethics* he presented a complete metaphysical system in austerely deductive form, proving theorems from axioms in the style of Euclid's geometry.

These excerpts begin with a fine expression of Spinoza's determination to treat human phenomena as subject to the same laws of nature as everything else—and therefore, in his rationalist viewpoint, capable of explanation by the deductive methods of mathematics. Then he declares his faith in the future progress of a science of human nature, suggesting very sagaciously that the physical complexities within the body greatly exceed what could then be conceived. This faith makes more intelligible his identification of mind and body as different aspects of one complex whole, and his statement that the mind has no power of its own to act independently of what is going on in the body.

Towards the end of the *Ethics,* Spinoza offers his own version of liberation, his prescription for how to achieve what he calls "blessedness," or freedom of mind. Anticipating Freud, he asserts that the more we can achieve adequate knowledge and understanding of our own emotional states, the more we can control them. (The term "affect" is often approximately translated as "emotion."

These extracts are from Part III, Preface and note to Proposition II, and Part V, Proposition IV; the translation is by W. H. White (Oxford University Press, 1910).

THE MATHEMATICAL STUDY OF HUMAN NATURE

Most persons who have written about the affects and man's conduct of life seem to discuss, not the natural things which follow the common laws of nature, but things which are outside her. They seem indeed to consider man in nature as a kingdom within a kingdom. For they believe that man disturbs rather than follows her order; that he has an absolute power over his own actions; and that he is altogether self-determined. They then proceed to attribute the cause of human weakness and changeableness, not to the common power of nature, but to some vice of human nature, which they therefore bewail, laugh at, mock, or, as is more generally the case, detest; whilst he who knows how to revile most eloquently or subtilly the weakness of the mind is looked upon as divine. It is true that very eminent men have not been wanting, to whose labour and industry we confess ourselves much indebted, who have written many excellent things about the right conduct of life, and who have given to mortals counsels full of prudence, but no one so far as I know has determined the nature and strength of the affects, and what the mind is able to do towards controlling them. I remember, indeed, that the celebrated Descartes, although he believed that the mind is absolute master over its own actions, tried nevertheless to explain by their first causes human affects, and at the same time to show the way by which the mind could obtain absolute power over them; but in my opinion he has shown nothing but the acuteness of his great intellect, as I shall make evident in the proper place, for I wish to return to those who prefer to detest and scoff at human affects and actions than understand them. To such as these it will doubtless seem a marvellous thing for me to endeavour to treat by a geometrical method the vices and follies of men, and to desire by a sure method to demonstrate those things which these people cry out against as being opposed to reason, or as being vanities, absurdities, and monstrosities. The following is my reason for so doing. Nothing happens in nature which can be attributed to any vice of nature, for she is always the same and everywhere one. Her virtue is the same, and her power of acting; that is to say, her laws and rules, according to which all things are and are changed from form to form, are everywhere and always the same; so that there must also be one and the same method of understanding the nature of all things whatsoever, that is to say, by the universal laws and rules of nature. The affects, therefore, of hatred, anger, envy, considered in themselves, follow from the same necessity and virtue of nature as other individual things; they have therefore certain causes through which they are to be understood, and certain properties which are just as worthy of being known as the properties of any other thing in the contemplation alone of which we delight. I shall, therefore, pursue the same method in considering the nature and strength of the affects and the power of the mind over them which I pursued in our previous discussion of God and the mind, and I shall consider human actions and appetites just as if I were considering lines, planes, or bodies. . . .

THE IDENTITY OF MIND AND BODY, AND THE ILLUSION OF FREEDOM

The mind and the body are one and the same thing, conceived at one time under the attribute of thought, and at another under that of extension. For this reason, the order or concatenation of things is one, whether nature be conceived under this or under that attribute, and consequently the order of the actions and passions of our body is coincident in nature with the order of the actions and passions of the mind.

Although these things are so, and no ground for doubting remains, I scarcely believe, nevertheless, that, without a proof derived from experience, men will be induced calmly to weigh what has been said, so firmly are they persuaded that, solely at the bidding of the mind, the body moves or rests, and does a number of things which depend upon the will of the mind alone, and upon the power of thought. For what the body can do no one has hitherto determined, that is to say, experience has taught no one hitherto what the body, without being determined by the mind, can do and what it cannot do from the laws of nature alone, in so far as nature is considered merely as corporeal. For no one as yet has understood the structure of the body so accurately as to be able to explain all its functions, not to mention the fact that many things are observed in brutes which far surpass human sagacity, and that sleep-walkers in their sleep do very many things which they dare not do when awake; all this showing

that the body itself can do many things from the laws of its own nature alone at which the mind belonging to that body is amazed.
Again, nobody knows by what means or by what method the mind
moves the body, nor how many degrees of motion it can communicate to the body, nor with what speed it can move the body. So that
it follows that when men say that this or that action of the body
springs from the mind which has command over the body, they do
not know what they say, and they do nothing but confess with pre-
tentious words that they know nothing about the cause of the action,
and see nothing in it to wonder at. But they will say, that whether
they know or do not know by what means the mind moves the
body, it is nevertheless in their experience that if the mind were not
fit for thinking the body would be inert. They say, again, it is in their
experience that the mind alone has power both to speak and be silent,
and to do many other things which they therefore think to be dependent on a decree of the mind. But with regard to the first assertion, I
ask them if experience does not also teach that if the body be sluggish
the mind at the same time is not fit for thinking? When the body is
asleep, the mind slumbers with it, and has not the power to think, as
it has when the body is awake. Again, I believe that all have discovered that the mind is not always equally fitted for thinking about
the same subject, but in proportion to the fitness of the body for this
or that image to be excited in it will the mind be better fitted to contemplate this or that object. But my opponents will say, that from
the laws of nature alone, in so far as it is considered to be corporeal
merely, it cannot be that the causes of architecture, painting, and
things of this sort, which are the results of human art alone, could be
deduced, and that the human body, unless it were determined and
guided by the mind, would not be able to build a temple. I have already shown, however, that they do not know what the body can
do, nor what can be deduced from the consideration of its nature
alone, and that they find that many things are done merely by the
laws of nature which they would never have believed to be possible
without the direction of the mind, as, for example, those things
which sleep-walkers do in their sleep, and at which they themselves
are astonished when they wake. I adduce also here the structure itself
of the human body, which so greatly surpasses in workmanship all

those things which are constructed by human art, not to mention what I have already proved, that an infinitude of things follows from nature under whatever attribute it may be considered.

With regard to the second point, I should say that human affairs would be much more happily conducted if it were equally in the power of men to be silent and to speak; but experience shows over and over again that there is nothing which men have less power over than the tongue, and that there is nothing which they are less able to do than to govern their appetites, so that many persons believe that we do those things only with freedom which we seek indifferently; as the desire for such things can easily be lessened by the recollection of another thing which we frequently call to mind; it being impossible, on the other hand, to do those things with freedom which we seek with such ardour that the recollection of another thing is unable to mitigate it. But if, however, we had not found out that we do many things which we afterwards repent, and that when agitated by conflicting affects we see that which is better and follow that which is worse, nothing would hinder us from believing that we do everything with freedom. Thus the infant believes that it is by free will that it seeks the breast; the angry boy believes that by free will he wishes vengeance; the timid man thinks it is with free will he seeks flight; the drunkard believes that by a free command of his mind he speaks the things which when sober he wishes he had left unsaid. Thus the madman, the chatterer, the boy, and others of the same kind, all believe that they speak by a free command of the mind, whilst, in truth, they have no power to restrain the impulse which they have to speak, so that experience itself, no less than reason, clearly teaches that men believe themselves to be free simply because they are conscious of their own actions, knowing nothing of the causes by which they are determined: it teaches, too, that the decrees of the mind are nothing but the appetites themselves, which differ, therefore, according to the different temper of the body. For every man determines all things from his affect; those who are agitated by contrary affects do not know what they want, whilst those who are agitated by no affect are easily driven hither and thither. All this plainly shows that the decree of the mind, the appetite, and determination of the body are coincident in nature, or rather that they are one and the same thing, which, when it is considered under the attribute of thought and manifested by that, is called a decree, and when it is considered under the attribute of extension and is deduced from the laws of motion and rest, is called a determination. This, however, will be better understood as we go on, for there is another thing which I wish to be observed here—that we cannot by a mental decree do a thing unless we recollect it. We cannot speak a word, for instance, unless we recollect it. But it is not in the free power of the mind either to recollect a thing or to forget it. It is believed, therefore, that the power of the mind extends only thus far—that from a mental decree we can speak or be silent about a thing only when we recollect it. But when we dream that we speak, we believe that we do so from a free decree of the mind; and yet we do not speak, or if we do, it is the result of a spontaneous motion of the body. We dream, again, that we are concealing things, and that we do this by virtue of a decree of the mind like that by which, when awake, we are silent about things we know. We dream, again, that, from a decree of the mind, we do some things which we should not dare to do when awake. And I should like to know, therefore, whether there are two kinds of decrees in the mind—one belonging to dreams and the other free. If this be too great nonsense, we must necessarily grant that this decree of the mind, which is believed to be free, is not distinguishable from the imagination or memory, and is nothing but the affirmation which the idea necessarily involves in so far as it is an idea.

These decrees of the mind, therefore, arise in the mind by the same necessity as the ideas of things actually existing. Consequently, those who believe that they speak, or are silent, or do anything else from a free decree of the mind, dream with their eyes open. . . .

THE LIBERATING POWER OF SELF-KNOWLEDGE

I pass at length to the other part of Ethic which concerns the method or way which leads to liberty. In this part, therefore, I shall treat of the power of reason, showing how much reason itself can control the affects, and then what is freedom of mind or blessedness. Thence we shall see how much stronger the wise man is than the ignorant. In what manner and in what way the intellect should be rendered perfect, and with what art the body is to be cared for in order that it may properly perform its functions, I have nothing to do with here; for the former belongs to logic, the latter to medicine. I shall occupy myself here, as I have said, solely with the power of the mind or of reason, first of all showing the extent and nature of the authority which it has over the affects in restraining them and governing them; for that we have not absolute authority over them we have already demonstrated. The Stoics indeed thought that the affects depend absolutely on our will, and that we are absolutely masters over them; but they were driven, by the contradiction of experience, though not by their own principles, to confess that not a little practice and study are required in order to restrain and govern the affects. This one of them attempted to illustrate, if I remember rightly, by the example of two dogs, one of a domestic and the other of a hunting breed; for he was able by habit to make the house-dog hunt, and the hunting dog, on the contrary, to desist from running after hares. . . .

As the power of the mind, as I have shown, is determined by intelligence alone, we shall determine by the knowledge of the mind alone the remedies against the affects—remedies which every one, I believe, has experienced, although there may not have been any accurate observation or distinct perception of them, and from this knowledge of the mind alone shall we deduce everything which relates to its blessedness. . . .

Since nothing exists from which some effect does not follow, and since we understand clearly and distinctly everything which follows from an idea which is adequate in us, it is a necessary consequence that every one has the power, partly at least, if not absolutely, of understanding clearly and distinctly himself and his affects, and consequently of bringing it to pass that he suffers less from them. We have therefore mainly to strive to acquire a clear and distinct knowledge as far as possible of each affect, so that the mind may be led to pass from the affect to think those things which it perceives clearly and distinctly, and with which it is entirely satisfied, and to strive also that the affect may be separated from the thought of an external cause and connected with true thoughts.

Thus not only love, hatred, etc., will be destroyed, but also the appetites or desires to which the affect gives rise cannot be excessive. For it is above everything to be observed that the appetite by which a man is said to act is one and the same appetite as that by which he is said to suffer. For example, we have shown that human nature is so constituted that every one desires that other people should live according to his way of thinking, a desire which in a man who is not guided by reason is a passion which is called ambition, and is not very different from pride; while, on the other hand, in a man who lives according to the dictates of reason it is an action or virtue which is called piety.

In the same manner, all the appetites or desires are passions only in so far as they arise from inadequate ideas, and are classed among the virtues whenever they are excited or begotten by adequate ideas; for all the desires by which we are determined to any action may arise either from adequate or inadequate ideas.

To return, therefore, to the point from which we set out: there is no remedy within our power which can be conceived more excellent for the affects than that which consists in a true knowledge of them, since the mind possesses no other power than that of thinking and forming adequate ideas, as we have shown.

Joseph Butler

Joseph Butler (1692–1752) was a bishop in the Church of England and also a considerable philosopher. In his *Sermons,* first published in 1726, he gave an extended analysis of how ethics depends on the *facts* about our human nature.

The English empirical approach exhibited in these excerpts is in marked contrast to Spinoza; yet there are elements in common. The influence of mechanistic science shows itself in Butler's example of the construction of a watch, but the notion of the parts being put together in such a way as to be adapted to a certain purpose goes beyond mechanics. No doubt, Butler was ready to interpret this notion in theological terms, but these do not appear here, so what he says is equally open to interpretation in terms of evolution (as the excerpt from K. Lorenz below may suggest). Butler's emphasis on the natural authority of what he rather vaguely calls conscience or reflection recalls Plato's reasoning part of the soul; the theme of reflection relates to that of self-knowledge in Spinoza, Freud, and other philosophers such as Sartre and Hampshire represented below.

In these paragraphs from the Preface to his *Sermons* (*The Works of Joseph Butler,* ed. W. E. Gladstone, Oxford University Press, 1897), Butler summarizes the main points of his argument.

HUMAN NATURE AS A WHOLE INTEGRATED BY REFLECTION

The two methods of handling morals, and their reciprocal service and respective merits

There are two ways in which the subject of morals may be treated. One begins from inquiring into the abstract relations of things: the other from a matter of fact, namely, what the particular nature of man is, its several parts, their economy or constitution; from whence it proceeds to determine what course of life it is, which is correspondent to this whole nature. In the former method the conclusion is expressed thus, that vice is contrary to the nature and reason of things: in the latter, that it is a violation or breaking in upon our own nature. Thus they both lead us to the same thing, our obligations to the practice of virtue; and thus they exceedingly strengthen and enforce each other. The first seems the most direct formal proof, and in some respects the least liable to cavil and dispute: the latter is in a peculiar manner adapted to satisfy a fair mind: and is more easily applicable to the several particular relations and circumstances in life.

The following Discourses proceed chiefly in this latter method. They were intended to explain what is meant by the nature of man, when it is said that virtue consists in following, and vice in deviating from it; and by explaining to shew that the assertion is true. . . .

A nature is an integer; its parts having reciprocal relations
needful to be known

Whoever thinks it worth while to consider this matter thoroughly, should begin with stating to himself exactly the idea of a system, economy, or constitution of any particular nature, or any particular thing: and he will, I suppose, find, that it is an one or a whole, made up of several parts; but yet, that the several parts even considered as a whole do not complete the idea, unless in the notion of a whole you include the relations and respects which those parts have to each other. Every work both of nature and of art is a system: and as every particular thing, both natural and artificial, is for some use or purpose out of and beyond itself, one may add, to what has been already brought into the idea of a system, its conduciveness to this one or more ends.

Illustrated by a watch

Let us instance in a watch—Suppose the several parts of it taken to pieces, and placed apart from each other: let a man have every so exact a notion of these several parts, unless he considers the respects and relations which they have to each other, he will not have any thing like the idea of a watch. Suppose these several parts brought together and any how united: neither will he yet, be the union ever so close, have an idea which will bear any resemblance to that of a watch. But let him view those several parts put together, or consider them as to be put together in the manner of a watch; let him form a notion of the relations which those several parts have to each other—all conducive in their respective ways to this purpose, showing the hour of the day; and then he has the idea of a watch.

In man the principal relation is that of conscience to the rest

Thus it is with regard to the inward frame of man. Appetites, passions, affections, and the principle of reflection, considered merely as the several parts of our inward nature, do not at all give us an idea of the system or constitution of this nature; because the constitution is formed by somewhat not yet taken into consideration, namely, by the relations which these several parts have to each other; the chief of which is the authority of reflection or conscience. It is from considering the relations which the several appetites and passions in the inward frame have to each other, and, above all, the supremacy of reflection or conscience, that we get the idea of the system or constitution of human nature.

Thus we find our nature adopted to virtue, as a watch to measuring time

And from the idea itself it will as fully appear, that this our nature, i.e. constitution, is adapted to virtue, as from the idea of a watch it appears, that its nature, i.e. constitution or system, is adapted to measure time. What in fact or event commonly happens is nothing to this question. Every work of art is apt to be out of order: but this is so far from being according to its system, that let the disorder increase, and it will totally destroy it. This is merely by way of explanation, what an economy, system, or constitution is. And thus far the cases are perfectly parallel. If we go further, there is indeed a difference, nothing to the present purpose, but too important an one ever to be omitted. A machine is inanimate and passive: but we are agents. Our constitution is put in our own power. We are charged with it; and therefore are accountable for any disorder or violation of it.

Vice, injustice, misery, all contrary to our nature:
variously in mode and degree

Thus nothing can possibly be more contrary to nature than vice; meaning by nature not only the *several parts* of our internal frame, but also the *constitution* of it. Poverty and disgrace, tortures and death, are not so contrary to it. Misery and injustice are indeed equally contrary to some different parts of our nature taken singly: but injustice is moreover contrary to the whole constitution of the nature.

If it be asked, whether this constitution be really what those philosophers meant, and whether they would have explained themselves in this manner; the answer is the same, as if it should be asked, whether a person, who had often used the word resentment, and felt the thing, would have explained this passion exactly in the same manner, in which it is done in one of these Discourses. As I have no doubt, but that this is a true account of that passion, which he referred to and intended to express by the word resentment; so I have no doubt, but that this is the true account of the ground of that conviction which they referred to, when they said, vice was contrary to nature. And though it should be thought that they meant no more than that vice was contrary to the higher and better part of our nature; even this implies such a constitution as I have endeavoured to explain. For the very terms, higher and better, imply a relation or respect of parts to each other; and these relative parts, being in one and the same nature, form a constitution, and are the very idea of it. They had a perception that injustice was contrary to their nature, and that pain was so also. They observed these two perceptions totally different, not in degree, but in kind: and the reflecting upon each of them, as they thus stood in their nature, wrought a full intuitive conviction, that more was due and of right belonged to one of these inward perceptions, than to the other; that it demanded in all cases to govern such a creature as man. So that, upon the whole, this is a fair and true account of what was the ground of their conviction; of what they intended to refer to, when they said, virtue consisted in following nature: a manner of speaking not loose and undeterminate, but clear and distinct, strictly just and true.

Men and brutes compared, as to instincts of action

Though I am persuaded the force of this conviction is felt by almost every one; yet since, considered as an argument and put in words, it appears somewhat abstruse, and since the connection of it is broken in the three first Sermons, it may not be amiss to give the reader the whole argument here in one view.

Mankind has various instincts and principles of action, as brute creatures have; some leading most directly and immediately to the good of the community, and some most directly to private good.

Man has several which brutes have not; particularly reflection or conscience, an approbation of some principles or actions, and disapprobation of others.

Brutes obey their instincts or principles of action, according to certain rules; suppose the constitution of their body, and the objects around them.

The generality of men obey these instincts in the mass, both the bad and the good

The generality of mankind also obey their instincts and principles, all of them; those propensions we call good, as well as the bad, according to the same rules; namely, the constitution of their body, and the external circumstances which they are in. Therefore it is not a true representation of mankind to affirm, that they are wholly governed by self-love, the love of power and sensual appetites: since, as on the one hand they are often actuated by these, without any regard to right or wrong; so on the other it is manifest fact, that the same persons, the generality, are frequently influenced by friendship, compassion, gratitude; and even a general abhorrence of what is base, and liking of what is fair and just, takes its turn amongst the other motives of action. This is the partial inadequate notion of human nature treated of in the first Discourse: and it is by this nature, if one may speak so, that the world is in fact influenced, and kept in that tolerable order, in which it is.

Brutes the like: but their nature has no discords

Brutes in acting according to the rules before mentioned, their bodily constitution and circumstances, act suitably to their whole nature. It is however to be distinctly noted, that the reason why we affirm this is not merely that brutes in fact act so; for this alone, however universal, does not at all determine, whether such course of action be correspondent to their whole nature: but the reason of the assertion is, that as in acting thus they plainly act conformably to somewhat in their nature, so, from all observations we are able to make upon them, there does not appear the least ground to imagine them to have any thing else in their nature, which requires a different rule or course of action.

Mankind also in acting thus would act suitably to their whole nature, if no more were to be said of man's nature than what has been now said; if that, as it is a true, were also a complete, adequate account of our nature.

But man has a ruling faculty of conscience or reflection

But that is not a complete account of man's nature. Somewhat further must be brought in to give us an adequate notion of it; namely, that one of those principles of action, conscience or re-flection, compared with the rest as they all stand together in the nature of man, plainly bears upon it marks of authority over all the rest, and claims the absolute direction of them all to allow or forbid their gratification: a disapprobation of reflection being in itself a principle manifestly superior to a mere propension. And the conclusion is, that to allow no more to this superior principle or part of our nature, than to other parts; to let it govern and guide only occasionally in common with the rest, as its turn happens to come, from the temper and circumstances one happens to be in; this is not to act conformably to the constitution of man: neither can any human creature be said to act conformably to his constitution of nature, unless he allows to that superior principle the absolute authority which is due to it. And this conclusion is abundantly confirmed from hence, that one may determine what course of action the economy of man's nature requires, without so much as knowing in what degrees of *strength* the several principles prevail, or which of them have actually the greatest influence.

Which claims to be his universal governor

The practical reason of insisting so much upon this natural authority of the principle of reflection or conscience is, that it seems in great measure overlooked by many, who are by no means the worse sort of men. It is thought sufficient to abstain from gross wickedness, and to be humane and kind to such as happen to come in their way. Whereas in reality the very constitution of our nature requires, that we bring our whole conduct before this superior faculty; wait its determination; enforce upon ourselves its authority, and make it the business of our lives, as it is absolutely the whole business of a moral agent, to conform ourselves to it. This is the true meaning of that ancient precept, *Reverence thyself.*

David Hume

The Scottish writer David Hume (1711–1776) is arguably the greatest philosopher the British Isles have produced. His magnum opus is the *Treatise of Human Nature,* published in three volumes in 1739–1740, from which we here reprint some excerpts from its Introduction. Spinoza, as the most thoroughgoing rationalist philosopher, tried to treat human nature by the methods of mathematics; Hume, in contrast, was the most thoroughgoing of the empiricists, and aimed to study human phenomena by the methods of physics (here called "Natural Philosophy").

Like so many thinkers, Hume took Newtonian mechanics as the model of a scientific theory. He hoped to find a few simple laws that could explain all the phenomena of human nature which we can observe in experience; for example, in the *Treatise,* I.1.iv, he talked of the principle of association of ideas playing the same role in the mental world as that of gravitational attraction in the natural world. But he did not expect such explanatory principles to have the self-evident quality that Spinoza required of his axioms; Hume's principles could be justified only by their fitting the data of human experience.

Hume was one of the first thinkers to work out in detail the consequences of such a fully naturalistic approach to understanding human nature. In this he was one of the foremost figures in the eighteenth-century European "Enlightenment"—that movement of thought which developed the hope that reason could be applied to understand human affairs and change them for the better. But it must be said that his actual treatment of human nature does not strictly follow the scientific methodology of his Introduction: his achievements are now seen as philosophy rather than science. We have to look to the nineteenth century for the real beginnings of social science.

It is hardly possible in an anthology of this kind to represent Hume's profound philosophy of human nature, which is expounded at length in the three volumes of the *Treatise.* But it is very much worthwhile extracting his famous attempt at reconciling those ancient philosophical warriors—determinism and free will. The second excerpt is from the version in Hume's *Enquiry Concerning Human Understanding* (Section VIII) which, together with his *Enquiry Concerning the Principles of Morals,* gives us a more popular and stylish exposition of the doctrines of his *Treatise*. The claim that there is no incompatibility between freedom, in the sense of the absence of external compulsion or threat, and determinism, in the sense of the existence of causes of action *within* the agent, has been a recurring feature of English empiricism (in Hobbes, Mill, and the twentieth century). That philosophers still find problems about freedom of action suggests that Hume's kind of reconciliation is not the end of the story, but it is surely a vital part of it.

THE POSSIBILITY OF AN EMPIRICAL SCIENCE
OF HUMAN NATURE

'Tis evident, that all the sciences have a relation, greater or less, to human nature; and that however wide any of them may seem to run from it, they still return back by one passage or another. Even *Mathematics, Natural Philosophy, and Natural Religion,* are in some measure dependent on the science of Man; since they lie under the cognizance of men, and are judged of by their powers and faculties. 'Tis impossible to tell what changes and improvements we might make in these sciences were we thoroughly acquainted with the extent and force of human understanding and cou'd explain the nature of the ideas we employ, and of the operations we perform in our reasonings. And these improvements are the more to be hoped for in natural religion, as it is not content with instructing us in the nature of superior powers, but carries its views farther, to their disposition towards us, and our duties towards them; and consequently we ourselves are not only the beings, that reason, but also one of the objects, concerning which we reason.

If therefore the sciences of Mathematics, Natural Philosophy, and Natural Religion, have such a dependence on the knowledge of man, what may be expected in the other sciences, whose connexion with human nature is more close and intimate? The sole end of logic is to explain the principles and operations of our reasoning faculty, and the nature of our ideas: morals and criticism regard our tastes and sentiments: and politics consider men as united in society, and dependent on each other. In these four sciences of *Logic, Morals, Criticism, and Politics,* is comprehended almost every thing, which it can any way import us to be acquainted with, or which can tend either to the improvement or ornament of the human mind.

Here then is the only expedient, from which we can hope for success in our philosophical researches, to leave the tedious lingring method, which we have hitherto followed, and instead of taking now and then a castle or village on the frontier, to march up directly to the capital or center of these sciences, to human nature itself; which being once masters of, we may every where else hope for an easy victory. From this station we may extend our conquests over all those sciences, which more intimately concern human life, and may afterwards proceed at leisure to discover more fully those which are the objects of pure curiosity. There is no question of importance, whose decision is not compriz'd in the science of man; and there is none, which can be decided with any certainty, before we become acquainted with that science. In pretending therefore to explain the principles of human nature, we in effect propose a compleat system of the sciences, built on a foundation almost entirely new, and the only one upon which they stand with any security.

And as the science of man is the only solid foundation for the other sciences, so the only solid foundation we can give to this science itself must be laid on experience and observation. . . .

For to me it seems evident, that the essence of the mind being equally unknown to us with that of external bodies, it must be equally impossible to form any notion of its powers and qualities otherwise than from careful and exact experiments, and the observation of those particular effects, which result from its different circumstances and situations. And tho' we must endeavour to render all our principles as universal as possible, by tracing up our experiments to the utmost, and explaining all effects from the simplest and fewest causes, 'tis still certain we cannot go beyond experience; and any hypothesis, that pretends to discover the ultimate qualities of human nature, ought at first to be rejected as presumptuous and chimerical.

I do not think a philosopher, who would apply himself so earnestly to the explaining of the ultimate principles of the soul, would show himself a great master in that very science of human nature, which he pretends to explain, or very knowing in what is naturally satisfactory to the mind of man. For nothing is more certain, than that despair has almost the same effect upon us with enjoyment, and that we are no sooner acquainted with the impossibility of satisfying any desire, than the desire itself vanishes. When we see, that we have arrived at the utmost extent of human reason, we sit down contented; tho' we be perfectly satisfied in the main of our ignorance, and perceive that we can give no reason for our most general and most refined principles, beside our experience of their reality; which is the reason of the mere vulgar, and what it required no study at first to have discovered for the most particular and most extraordinary phenomenon. And as this impossibility of making any farther progress is enough to satisfy the reader, so the writer may derive a more delicate satisfaction from the free confession of his ignorance, and from his prudence in avoiding that error into which so many have fallen of imposing their conjectures and hypotheses on the world for the most certain principles. When this mutual contentment and satisfaction can be obtained betwixt the master and scholar, I know not what more we can require of our philosophy.

But if this impossibility of explaining ultimate principles should be esteemed a defect in the science of man, I will venture to affirm, that 'tis a defect common to it with all the sciences, and all the arts, in which we can employ ourselves, whether they be such as are cultivated in the schools of the philosophers, or practised in the shops of the meanest artizans. None of them can go beyond experience, or establish any principles which are not founded on that authority. Moral philosophy has, indeed, this peculiar disadvantage, which is not found in natural, that in collecting its experiments, it cannot make them purposely, with premeditation, and after such a manner as to satisfy itself concerning every particular difficulty which may arise. When I am at a loss to know the effects of one body upon another in any situation, I need only put them in that situation, and observe what results from it. But should I endeavour to clear up after the same manner any doubt in moral philosophy, by placing myself in the same case with that which I consider, 'tis evident this reflection and premeditation would so disturb the operation of my natural principles, as must render it impossible to form any just conclusion from the phenomenon. We must therefore glean up our experiments in this science from a cautious observation of human life, and take them as they appear in the common course of the world, by men's behaviour in company, in affairs, and in their pleasures. Where experiments of this kind are judiciously collected and compared, we may hope to establish on them a science which will not be inferior in certainty, and will be much superior in utility to any other of human comprehension.

THE COMPATABILITY OF DETERMINISM AND FREEDOM

I hope to make it appear that all men have ever agreed in the doctrine both of necessity and of liberty, according to any reasonable sense, which can be put on these terms; and that whole controversy has hitherto turned merely upon words. We shall begin with examining the doctrine of necessity.

It is universally allowed that matter, in all its operations, is actuated by a necessary force, and that every natural effect is so precisely determined by the energy of its cause that no other effect, in such particular circumstances, could possibly have resulted from it. The degree and direction of every motion is, by the laws of nature, prescribed with such exactness that a living creature may as soon arise from the shock of two bodies as motion in any other degree or direction than what is actually produced by it. Would we, therefore, form a just and precise idea of *necessity,* we must consider whence that idea arises when we apply it to the operation of bodies.

It seems evident that, if all the scenes of nature were continually shifted in such a manner that no two events bore any resemblance to each other, but every object was entirely new without any similitude to whatever had been seen before, we should never, in that case, have attained the least idea of necessity, or of a connexion among these objects. We might say, upon such a supposition, that one object or event has followed another; not that one was produced by the other. The relation of cause and effect must be utterly unknown to mankind. Inference and reasoning concerning the operations of nature would, from that moment, be at an end; and the memory and senses remain the only canals, by which the knowledge of any real existence could possibly have access to the mind. Our idea, therefore, of necessity and causation arises entirely from the uniformity observable in the operations of nature, where similar objects are constantly conjoined together, and the mind is determined by custom to infer the one from the appearance of the other. These two circumstances form the whole of that necessity, which we ascribe to matter. Beyond the constant *conjunction* of similar objects, and the consequent *inference* from one to the other, we have no notion of any necessity or connexion.

If it appear, therefore, that all mankind have ever allowed, without any doubt or hesitation, that these two circumstances take place in the voluntary actions of men, and in the operations of mind; it must follow, that all mankind have ever agreed in the doctrine of necessity, and that they have hitherto disputed, merely for not understanding each other.

As to the first circumstance, the constant and regular conjunction of similar events, we may possibly satisfy ourselves by the following considerations. It is universally acknowledged that there is a great uniformity among the actions of men, in all nations and ages, and that human nature remains still the same, in its principles and operations. The same motives always produce the same actions: The same events follow from the same causes. Ambition, avarice, self-love, vanity, friendship, generosity, public spirit: these passions, mixed in various degrees, and distributed through society, have been, from the beginning of the world, and still are, the source of all the actions and enterprises, which have ever been observed among mankind. Would you know the sentiments, inclinations, and course of life of the Greeks and Romans? Study well the temper and actions of the French and English: You cannot be much mistaken in transferring to the former *most* of the observations which you have made with regard to the latter. Mankind are so much the same, in all times and places, that history informs us of nothing new or strange in this particular. Its chief use is only to discover the constant and universal principles of human nature, by showing men in all varieties of circumstances and situations, and furnishing us with materials from which we may form our observations and become acquainted with the regular springs of human action and behaviour. These records of wars, intrigues, factions, and revolutions, are so many collections of experiments, by which the politician or moral philosopher fixes the principles of his science, in the same manner as the physician or natural philosopher becomes acquainted with the nature of plants, minerals, and other external objects, by the experiments which he forms concerning them. Nor are the earth, water, and other elements, examined by Aristotle, and Hippocrates, more like to those which at present lie under our observation than the men described by Polybius and Tacitus are to those who now govern the world.

Should a traveller, returning from a far country, bring us an account of men, wholly different from any with whom we were ever acquainted; men, who were entirely divested of avarice, ambition, or revenge; who knew no pleasure but friendship, generosity, and public spirit; we should immediately, from these circumstances, detect the falsehood, and prove him a liar, with the same certainty as if he had stuffed his narration with stories of centaurs and dragons, miracles and prodigies. And if we would explode any forgery in history, we cannot make use of a more convincing argument, than to prove, that the actions ascribed to any person are directly contrary to the course of nature, and that no human motives, in such circumstances, could ever induce him to such a conduct. The veracity of Quintus Curtius is as much to be suspected, when he describes the supernatural courage of Alexander, by which he was hurried on singly to attack multitudes, as when he describes his supernatural force and activity, by which he was able to resist them. So readily and universally do we acknowledge a uniformity in human motives and actions as well as in the operations of body. . . .

Thus it appears, not only that the conjunction between motives and voluntary actions is as regular and uniform as that between the cause and effect in any part of nature; but also that this regular conjunction has been universally acknowledged among mankind, and has never been the subject of dispute either in philosophy or common life. Now, as it is from past experience that we draw all inferences concerning the future, and as we conclude that objects will always be conjoined together which we find to have always been conjoined; it may seem superfluous to prove that this experienced uniformity in human actions is a source whence we draw *inferences* concerning them. But in order to throw the argument into a greater variety of lights we shall also insist, though briefly, on this latter topic. . . .

And indeed, when we consider how aptly *natural* and *moral* evidence link together, and form only one chain of argument, we shall make no scruple to allow that they are of the same nature, and derived from the same principles. A prisoner who has neither money nor interest, discovers the impossibility of his escape, as well when he considers the obstinacy of the gaoler, as the walls and bars with which he is surrounded; and, in all attempts for his freedom, chooses rather to work upon the stone and iron of the one, than upon the inflexible nature of the other. The same prisoner, when conducted to the scaffold, foresees his death as certainly from the constancy and fidelity of his guards, as from the operation of the axe or wheel. His mind runs along a certain train of ideas: The refusal of the soldiers to consent to his escape; the action of the executioner; the separation of the head and body; bleeding, convulsive motions, and death. Here is a connected chain of natural causes and voluntary actions; but the mind feels no difference between them in passing from one link to another: Nor is less certain of the future event than if it were connected with the objects present to the memory or senses, by a train of causes, cemented together by what we are pleased to call a *physical* necessity. The same experienced union has the same effect on the mind, whether the united objects be motives, volition, and actions; or figure and motion. We may change the name of things; but their nature and their operation on the understanding never change.

Were a man, whom I know to be honest and opulent, and with whom I live in intimate friendship, to come into my house, where I am surrounded with my servants, I rest assured that he is not to stab me before he leaves it in order to rob me of my silver standish; and I no more suspect this event than the falling of the house itself, which is new, and solidly built and founded.—*But he may have been seized with a sudden and unknown frenzy.*—So may a sudden earthquake arise, and shake and tumble my house about my ears. I shall therefore change the suppositions. I shall say that I know with certainty that he is not to put his hand into the fire and hold it there till it be consumed: And this event, I think I can foretell with the same assurance, as that, if he throw himself out at the window, and meet with no obstruction, he will not remain a moment suspended in the air. No suspicion of an unknown frenzy can give the least possibility to the former event, which is so contrary to all the known principles of human nature. A man who at noon leaves his purse full of gold on the pavement at Charing-Cross, may as well expect that it will fly away like a feather, as that he will find it untouched an hour after. Above one half of human reasonings contain inferences of a similar nature, attended with more or less degrees of certainty proportioned to our experience of the usual conduct of mankind in such particular situations. . . .

But to proceed in this reconciling project with regard to the question of liberty and necessity; the most contentious question of metaphysics, the most contentious science; it will not require many words to prove, that all mankind have ever agreed in the doctrine of liberty as well as in that of necessity, and that the whole dispute, in this respect also, has been hitherto merely verbal. For what is meant by liberty, when applied to voluntary actions? We cannot surely mean that actions have so little connexion with motives, inclinations, and circumstances, that one does not follow with a certain degree of uniformity from the other, and that one affords no inference by which we can conclude the existence of the other. For these are plain and acknowledged matters of fact. By liberty, then, *we can only mean a power of acting or not acting, according to the determinations of the will;* that is, if we choose to remain at rest, we may; if we choose to move, we also may. Now this hypothetical liberty is universally allowed to belong to every one who is not a prisoner and in chains. Here, then, is no subject of dispute.

Whatever definition we may give of liberty, we should be careful to observe two requisite circumstances; *first,* that it be consistent with plain matter of fact; *secondly,* that it be consistent with itself. If we observe these circumstances, and render our definition intelligible, I am persuaded that all mankind will be found of one opinion with regard to it.

It is universally allowed that nothing exists without a cause of its existence, and that chance, when strictly examined, is a mere negative word, and means not any real power which has anywhere a being in nature. But it is pretended that some causes are necessary, some not necessary. Here then is the advantage of definitions. Let any one *define* a cause, without comprehending, as a part of the definition, a *necessary connexion* with its effect; and let him show distinctly the origin of the idea, expressed by the definition; and I shall readily give up the whole controversy. But if the foregoing explication of the matter be received, this must be absolutely impracticable. Had not objects a regular conjunction with each other, we should never have entertained any notion of cause and effect; and this regular conjunction produces that inference of the understanding, which is the only connexion, that we can have any comprehension of. Whoever attempts a definition of cause, exclusive of these circumstances, will be obliged either to employ unintelligible terms or such as are synonymous to the term which he endeavours to define. And if the definition above mentioned be admitted; liberty, when opposed to necessity, not to constraint, is the same thing with chance; which is universally allowed to have no existence.

There is no method of reasoning more common, and yet none more blameable, than, in philosophical disputes, to endeavour the refutation of any hypothesis, by a pretence of its dangerous consequences to religion and morality. When any opinion leads to absurdities, it is certainly false; but it is not certain that an opinion is false, because it is of dangerous consequence. Such topics, therefore, ought entirely to be forborne; as serving nothing to the discovery of truth but only to make the person of an antagonist odious. This I observe in general, without pretending to draw any advantage from it. I frankly submit to an examination of this kind, and shall venture to affirm that the doctrines, both of necessity and of liberty, as above explained, are not only consistent with morality, but are absolutely essential to its support.

Necessity may be defined two ways, conformably to the two definitions of *cause,* of which it makes an essential part. It consists either in the constant conjunction of like objects, or in the inference of the understanding from one object to another. Now necessity, in both these senses, (which, indeed, are at bottom the same) has universally, though tacitly, in the schools, in the pulpit, and in common life, been allowed to belong to the will of man; and no one has ever pretended to deny that we can draw inferences concerning human actions, and that those inferences are founded on the experienced union of like actions, with like motives, inclinations, and circumstances. The only particular in which any one can differ, is, that either, perhaps, he will refuse to give the name of necessity to this property of human actions: But as long as the meaning is under-stood, I hope the word can do no harm: Or that he will maintain it possible to discover something farther in the operations of matter. But this, it must be acknowledged, can be of no consequence to morality or religion, whatever it may be to natural philosophy or metaphysics. We may here be mistaken in asserting that there is no idea of any other necessity or connexion in the actions of body: But surely we ascribe nothing to the actions of the mind, but what every-one does, and must readily allow of. We change no circumstance in the received orthodox system with regard to the will, but only in that with regard to material objects and causes. Nothing, therefore, can be more innocent, at least, than this doctrine.

All laws being founded on rewards and punishments, it is sup-posed as a fundamental principle, that these motives have a regular and uniform influence on the mind, and both produce the good and prevent the evil actions. We may give to this influence what name we please; but, as it is usually conjoined with the action, it must be esteemed a *cause,* and be looked upon as an instance of that neces-sity, which we would here establish.

The only proper object of hatred or vengeance is a person or creature, endowed with thought and consciousness; and when any criminal or injurious actions excite that passion, it is only by their relation to the person, or connexion with him. Actions are, by their very nature, temporary and perishing; and where they proceed not for some *cause* in character and disposition of the person who per-formed them, they can neither redound to his honour, if good; nor infamy, if evil. The actions themselves may be blameable; they may be contrary to all the rules of morality and religion: But the person is not answerable for them; and as they proceed from nothing in him that is durable and constant, and leave nothing of that nature behind them, it is impossible he can, upon their account, become the object of punishment or vengeance. According to the principle, therefore, which denies necessity, and consequently causes, a man is as pure and untainted, after having committed the most horrid crime, as at the first moment of his birth, nor is his character anywise concerned in his actions, since they are not derived from it, and the wickedness of the one can never be used as a proof of the depravity of the other.

Men are not blamed for such actions as they perform ignorantly and casually, whatever may be the consequences. Why? but because the principles of these actions are only momentary, and terminate in them alone. Men are less blamed for such actions as they perform hastily and unpremeditately than for such as proceed from deliberation. For what reason? but because a hasty temper, though a constant cause or principle in the mind, operates only by intervals, and infects not the whole character. Again, repentance wipes off every crime, if attended with a reformation of life and manners. How is this to be accounted for? but by asserting that actions render a person criminal merely as they are proofs of criminal principles in the mind; and when, by an alteration of these principles, they cease to be just proofs, they likewise cease to be criminal. But, except upon the doctrine of necessity, they never were just proofs, and consequently never were criminal.

It will be equally easy to prove, and from the same arguments, that *liberty,* according to that definition above mentioned, in which all men agree, is also essential to morality, and that no human actions, where it is wanting, are susceptible of any moral qualities, or can be the objects either of approbation or dislike. For as actions are objects of our moral sentiment, so far only as they are indications of the internal character, passions, and affections; it is impossible that they can give rise either to praise or blame, where they proceed not from these principles, but are derived altogether from external violence.

John Stuart Mill

John Stuart Mill (1806–1873), writing a century after Hume, continues the British empiricist tradition and expresses similar hopes for a science of human nature. In these extracts from *A System of Logic* (1st ed., 1843) Book VI, Chapters III, IV, and VI, we find slightly greater sophistication about what forms such scientific study might take. He makes a useful distinction between exact sciences and inexact ones, such as tidology and meteorology, and claims that the social sciences are of the latter kind. In the second excerpt here, Mill offers reasons for thinking that there are "laws of Mind"—by which he means causal laws relating introspectable mental states—which we cannot as yet derive from the laws of physiology, so that psychology remains a separate science with its own usefulness. He recognizes the greater quantitative difficulties in the way of a science of human *societies,* but sees no difference of principle.

THAT THERE IS, OR MAY BE,
AN INEXACT SCIENCE OF HUMAN NATURE

§ 1.   It is a common notion, or at least it is implied in many common modes of speech, that the thoughts, feelings, and actions of sentient beings are not a subject of science, in the same strict sense in which this is true of the objects of outward nature. This notion seems to involve some confusion of ideas, which it is necessary to begin by clearing up.

Any facts are fitted, in themselves, to be a subject of science, which follow one another according to constant laws; although those laws may not have been discovered, nor even be discoverable by our existing resources. Take, for instance, the most familiar class of meteorological phenomena, those of rain and sunshine. Scientific inquiry has not yet succeeded in ascertaining the order of antecedence and consequence among these phenomena, so as to be able, at least in our regions of the earth, to predict them with certainty, or even with any high degree of probability. Yet no one doubts that the phenomena depend on laws, and that these must be derivative laws resulting from known ultimate laws, those of heat, vaporization, and elastic fluids. Nor can it be doubted that if we were acquainted with all the antecedent circumstances, we could, even from those more general laws, predict (saving difficulties of calculation) the state of the weather at any future time. Meteorology, therefore, not only has in itself every natural requisite for being, but actually is, a science; though, from the difficulty of observing the facts on which the phenomena depend (a difficulty inherent in the peculiar nature of those phenomena) the science is extremely imperfect; and were it perfect, might probably be of little avail in practice, since the data requisite for applying its principles to particular instances would rarely be procurable.

A case may be conceived, of an intermediate character between the perfection of science, and this its extreme imperfection. It may happen that the greater causes, those on which the principle part of the phenomenon depends, are within reach of observation and measurement; so that if no other causes intervened, a complete explanation could be given not only of the phenomenon in general, but of all the variations and modifications which it admits of. But inasmuch as other, perhaps many other causes, separately insignificant in their effects, co-operate or conflict in many or in all cases with those greater causes; the effect, accordingly, presents more or less of aberration from what would be produced by the greater causes alone. Now if these minor causes are not so constantly accessible, or not accessible at all, to accurate observation; the principal mass of the effect may still, as before, be accounted for, and even predicted; but there will be variations and modifications which we shall not be competent to explain thoroughly, and our predictions will not be fulfilled accurately, but only approximately.

It is thus, for example, with the theory of the tides. No one doubts that Tidology (as Dr. Whewell proposes to call it) is really a science. As much of the phenomena as depends on the attraction of the sun and moon is completely understood, and may in any, even unknown, part of the earth's surface, be foretold with certainty; and the far greater part of the phenomena depends on those causes. But circumstances of a local or casual nature, such as the configuration of the bottom of the ocean, the degree of confinement from shores, the direction of the wind, etc., influence, in many or in all places, the height and time of the tide; and a portion of these circumstances being either not accurately knowable, not precisely measurable, or at least not capable of being certainly foreseen, the tide in known places commonly varies from the calculated result of general principles by some difference that we cannot explain, and in unknown ones may vary from it by a difference that we are not able to foresee or conjecture. Nevertheless, not only is it certain that these variations depend on causes, and follow their causes by laws of unerring uniformity; not only, therefore, is tidology a science, like meteorology, but it is, what hitherto at least meteorology is not, a science largely available in practice. General laws may be laid down respecting the tides, predictions may be founded on those laws, and the result will in the main, though often not with complete accuracy, correspond to the predictions.

And this is what is or ought to be meant by those who speak of sciences which are not *exact* sciences. Astronomy was once a science, without being an exact science. It could not become exact until not only the general course of the planetary motions, but the perturbations also, were accounted for, and referred to their causes. It has become an exact science, because its phenomena have been brought under laws comprehending the whole of the causes by which the phenomena are influenced, whether in a great or only in a trifling degree, whether in all or only in some cases, and assigning to each of those causes the share of effect which really belongs to it. But in the theory of the tides the only laws as yet accurately ascertained are those of the causes which affect the phenomenon in all cases, and in a considerable degree; while others which affect it in some cases only, or, if in all, only in a slight degree, have not been sufficiently ascertained and studied to enable us to lay down their laws; still less to deduce the completed law of the phenomenon, by compounding the effects of the greater with those of the minor causes. Tidology, therefore, is not yet an exact science; not from any inherent incapacity of being so, but from the difficulty of ascertaining with complete precision the real derivative uniformities. By combining, however, the exact laws of the greater causes, and of such of the minor ones as are sufficiently known, with such empirical laws or such approximate generalizations respecting the miscellaneous variations as can be obtained by specific observation, we can lay down general propositions which will be true in the main, and on which, with allowance for the degree of their probable inaccuracy, we may safely ground our expectations and our conduct.

§ 2.   The science of human nature is of this description. It falls far short of the standard of exactness now realized in Astronomy; but there is no reason that it should not be as much a science as Tidology is, or as Astronomy was when its calculations had only mastered the main phenomena, but not the perturbations.

The phenomena with which this science is conversant being the thoughts, feelings, and actions of human beings, it would have attained the ideal perfection of a science if it enabled us to foretell how an individual would think, feel, or act, throughout life, with the same certainty with which astronomy enables us to predict the places and the occultations of the heavenly bodies. It needs scarcely be stated that nothing approaching to this can be done. The actions of individuals could not be predicted with scientific accuracy, were it only because we cannot foresee the whole of the circumstances in which those individuals will be placed. But further, even in any combination of (present) circumstances, no assertion, which is both precise and universally true, can be made respecting the manner in which human beings will think, feel, or act. This is not, however, because every person's modes of thinking, feeling, and acting, do not depend on causes; nor can we doubt that if, in the case of any individual, our data could be complete, we even now know enough of the ultimate laws by which mental phenomena are determined, to enable us in many cases to predict, with tolerable certainty, what, in the greater number of supposable combinations of circumstances, his conduct or sentiments would be. But the impressions and actions of human beings are not solely the result of their present circumstances, but the joint result of those circumstances and of the characters of the individuals: and the agencies which determine human character are so numerous and diversified, (nothing which has happened to the person throughout life being without its portion of influence,) that in the aggregate they are never in any two cases exactly similar. Hence, even if our science of human nature were theoretically perfect, that is, if we could calculate any character as we can calculate the orbit of any planet, *from given data;* still, as the data are never all given, nor ever precisely alike in different cases, we could neither make positive predictions, nor lay down universal propositions.

Inasmuch, however, as many of those effects which it is of most importance to render amenable to human foresight and control, are determined, like the tides, in an incomparably greater degree by general causes, than by all partial causes taken together; depending in the main on those circumstances and qualities which are common to all mankind, or at least to large bodies of them, and only in a small degree on the idiosyncrasies of organization or the peculiar history of individuals; it is evidently possible with regard to all such effects, to make predictions which will *almost* always be verified, and general propositions which are almost always true. And whenever it is sufficient to know how the great majority of the human race, or of some nation or class of persons, will think, feel, and act, these propositions are equivalent to universal ones. For the purposes of political and social science this is sufficient. As we formerly remarked, an approximate generalization is, in social inquiries, for most practical purposes equivalent to an exact one; that which is only probable when asserted of individual human beings indiscriminately selected, being certain when affirmed of the character and collective conduct of masses.

It is no disparagement, therefore, to the science of Human Nature, that those of its general propositions which descend sufficiently into detail to serve as a foundation for predicting phenomena in the concrete, are for the most part only approximately true. But in order to give a genuinely scientific character to the study, it is indispensible that these approximate generalizations, which in themselves would amount only to the lowest kind of empirical laws, should be connected deductively with the laws of nature from which they result; should be resolved into the properties of the causes on which the phenomena depend. In other words, the science of Human Nature may be said to exist, in proportion as the approximate truths, which compose a practical knowledge of mankind, can be exhibited as corollaries from the universal laws of human nature on which they rest; whereby the proper limits of those approximate truths would be shown, and we should be enabled to deduce others for any new state of circumstances, in anticipation of specific experience. . . .

PSYCHOLOGY AS STUDY OF THE LAWS OF MIND

§ 1.   What the Mind is, as well as what Matter is, or any other question respecting Things in themselves, as distinguished from their sensible manifestations, it would be foreign to the purpose of this treatise to consider. Here, as throughout our inquiry, we shall keep clear of all speculations respecting the mind's own nature, and shall understand by the laws of mind, those of mental Phenomena; of the various feelings or states of consciousness of sentient beings. These, according to the classification we have uniformly followed, consist of Thoughts, Emotions, Volitions, and Sensations: the last being as truly states of Mind as the three former. It is usual indeed to speak of sensations as states of body, not of mind. But this is the common confusion, of giving one and the same name to a phenomenon and to the proximate cause or conditions of the phenomenon. The immediate antecedent of a sensation is a state of body, but the sensation itself is a state of mind. If the word mind means anything, it means that which feels. Whatever opinion we hold respecting the fundamental identity or diversity of matter and mind, in any case the distinction between mental and physical facts, between the internal and the external world, will always remain, as a matter of classification: and in that classification, sensations, like all other feelings, must be ranked as mental phenomena. The mechanism of their production, both in the body itself and in what is called outward nature, is all that can with any propriety be classed as physical.

The phenomena of mind, then, are the various feelings of our nature, both these improperly called physical, and those peculiarly designated as mental: and by the laws of mind, I mean the laws according to which those feelings generate one another.

§ 2.   All states of mind are immediately caused either by other states of mind, or by states of body. When a state of mind is produced by a state of mind, I call the law concerned in the case, a law of Mind. When a state of mind is produced directly by a state of body, the law is a law of Body, and belongs to physical science.

With regard to those states of mind which are called sensations, all are agreed that these have for their immediate antecedents, states of body. Every sensation has for its proximate cause some affection of the portion of our frame called the nervous system; whether this affection originates in the action of some external object, or in some pathological condition of the nervous organization itself. The laws of this portion of our nature—the varieties of our sensations, and the physical conditions on which they proximately depend—manifestly belong to the province of Physiology.

Whether the remainder of our mental states are similarly dependent on physical conditions, is one of the *vexutae questiones* in the science of human nature. It is still disputed whether our thoughts, emotions, and volitions are generated through the intervention of material mechanism; whether we have organs of thought and of emotion, in the same sense in which we have organs of sensation. Many eminent physiologists hold the affirmative. These contend that a thought (for example) is as much the result of nervous agency, as a sensation: that some particular state of our nervous system, in particular of that central portion of it called the brain, invariably precedes, and is presupposed by, every state of our consciousness. According to this theory, one state of mind is never really produced by another: all are produced by states of body. When one thought seems to call up another by association, it is not really a thought which [recalls] a thought; the association did not exist between the two thoughts, but between the two states of the brain or nerves which preceded the thoughts: one of those states [recalls] the other, each being attended, in its passage, by the particular state of consciousness which is consequent on it. On this theory of uniformities of succession among states of mind would be mere derivative uniformities, resulting from the laws of succession of the bodily states which cause them. There would be no original mental laws, no Laws of Mind in the sense in which I use the term, at all: and mental science would be a mere branch, though the highest and most recondite branch, of the science of physiology. M. Comte, accordingly, claims the scientific cognizance of moral and intellectual phenomena exclusively for physiologists; and not only denies to Psychology, or Mental Philosophy properly so called, the character of a science, but places it, in the chimerical nature of its objects and pretensions, almost on a par with astrology.

But, after all has been said which can be said, it remains incontestable that there exist uniformities of succession among states of mind, and that these can be ascertained by observation and experiment. Further, that every mental state has a nervous state for its immediate antecedent and proximate cause, though extremely probable, cannot hitherto be said to be proved, in the conclusive manner in which this can be proved of sensations; and even were it certain, yet every one must admit that we are wholly ignorant of the characteristics of these nervous states; we know not, and at present have not means of knowing, in what respect one of them differs from another; and our only mode of studying their successions or coexistences must be by observing the successions and coexistences of the mental states, of which they are supposed to be the generators or causes. The successions, therefore, which obtain among mental phenomena, do not admit of being deduced from the physiological laws of our nervous organization: and all real knowledge of them must continue, for a long time at least, if not for ever, to be sought in the direct study, by observation and experiment, of the mental successions themselves. Since therefore the order of our mental phenomena must be studied in those phenomena, and not inferred from the laws of any phenomena more general, there is a distinct and separate Science of Mind.

The relations, indeed, of that science to the science of physiology must never be overlooked or undervalued. It must by no means be forgotten that the laws of mind may be derivative laws resulting from the laws of animal life, and that their truth therefore may ultimately depend on physical conditions; and the influence of physiological states or physiological changes in altering or counteracting the mental successions, is one of the most important departments of psychological study. But, on the other hand, to reject the resource of psychological analysis, and construct the theory of the mind solely on such data as physiology at present affords, seems to me as great an error in principle, and an even more serious one in practice. Imperfect as is the science of mind, I do not scruple to affirm, that is in a considerably more advanced state than the portion of physiology which corresponds to it; and to discard the former for the latter appears to me an infringement of the true canons of inductive philosophy, which must produce, and which does produce, erroneous conclusions in some very important departments of the science of human nature.

§ 3.   The subject, then, of Psychology, is the uniformities of succession, the laws, whether ultimate or derivative, according to which one mental state succeeds another; is caused by, or at least, is caused to follow, another. . . .

THE POSSIBILITY OF SOCIAL SCIENCE

§ 1.   Next after the science of individual man, comes the science of man in society: of the actions of collective masses of mankind, and the various phenomena which constitute social life.

If the formation of individual character is already a complex subject of study, this subject must be, in appearance at least, still more complex; because the number of concurrent causes, all exercising more or less influence on the total effect, is greater, in the proportion in which a nation, or the species at large, exposes a larger surface to the operation of agents, psychological and physical, than any single individual. If it was necessary to prove, in opposition to an existing prejudice, that the simpler of the two is capable of being a subject of science; the prejudice is likely to be yet stronger against the possibility of giving a scientific character to the study of Politics, and of the phenomena of Society. It is, accordingly, but of yesterday that the conception of a political or social science has existed, anywhere but in the mind of here and there an insulated thinker, generally very ill prepared for its realization; though the subject itself has of all others engaged the most general attention, and been a theme of interested and earnest discussions, almost from the beginning of recorded time. . . .

§ 2.   All phenomena of society are phenomena of human nature, generated by the action of outward circumstances upon masses of human beings: and if, therefore, the phenomena of human thought, feeling, and action, are subject to fixed laws, the phenomena of society cannot but conform to fixed laws, the consequence of the preceding. There is, indeed, no hope that these laws, though our knowledge of them were as certain and as complete as it is in astronomy, would enable us to predict the history of society, like that of the celestial appearances, for thousands of years to come. But the difference of certainty is not in the laws themselves, it is in the data to which these laws are to be applied. In astronomy the causes influencing the result are few, and change little, and that little according to known laws; we can ascertain what they are now, and thence determine what they will be at an epoch of a distant future. The data, therefore, in astronomy, are as certain as the laws themselves. The circumstances, on the contrary, which influence the condition and progress of society, are innumerable, and perpetually changing; and though they all change in obedience to causes, and therefore to laws, the multitude of the causes is so great as to defy our limited powers of calculation. Not to say that the impossibility of applying precise numbers to facts of such a description, would set an impassable limit to the possibility of calculating them beforehand, even if the powers of the human intellect were otherwise adequate to the task.

But, as before remarked, an amount of knowledge quite insufficient for prediction, may be most valuable for guidance. The science of society would have attained a very high point of perfection, if it enabled us, in any given condition of social affairs, in the condition for instance of Europe or any European country at the present time, to understand by what causes it had, in any and every particular, been made what it was; whether it was tending to any, and to what, changes; what effects each feature of its existing state was likely to produce in the future; and by what means any of those effects might be prevented, modified, or accelerated, or a different class of effects superinduced. There is nothing chimerical in the hope that general laws, sufficient to enable us to answer these various questions for any country or time with the individual circumstances of which we are well acquainted, do really admit of being ascertained; and that the other branches of human knowledge, which this undertaking presupposes, are so far advanced that the time is ripe for its commencement. Such is the object of the Social Science.

Émile Durkheim

It seems appropriate to place these brief excerpts from Durkheim next, since they can be seen as continuing the argument about the methodology of social science and as introducing themes developed by Marx and Freud. The Frenchman Émile Durkheim (1858–1917) is, along with Marx, Comte, and Weber, one of the founding fathers of sociology. In these excerpts he defines what is to count as a social fact, that is, as a datum for sociology to explain. He believes that facts about societies are irreducible to psychological facts about individuals, let alone to physical facts. So he would agree with Mill's view that psychology cannot be derived from physiology, and he would add that sociology is equally irreducible to psychology. Thus, we would have at least three different kinds of scientific study of human nature.

In another respect Durkheim represents a new level of sophistication in dealing with the nature of social science—in his recognition of the difficulties in *knowing* these different levels of social and psychological fact. He asserts that our individual consciousness does not give us any adequate understanding of psychological facts about ourselves, and that knowledge of social facts is even less guaranteed just by our being members of the relevant society. These points are fundamental to the work of Freud and Marx (though, because of the date, the "objective psychology" to which Durkheim refers can hardly be that of Freud).

The excerpts here are from the Preface to his *Rules of Sociological Method* (1st ed., 1895, translated by A. Giddens, in Durkheim, *Selected Writings,* Cambridge University Press, 1972.)

THE IRREDUCIBILITY OF SOCIAL FACTS, AND THEIR UNAVAILABILITY TO CONSCIOUSNESS

When I carry out my obligations as brother, husband, or citizen, when I comply with contracts, I perform duties which are defined, externally to myself and my acts, in law and in custom. Even if they conform to my own sentiments and I feel their reality subjectively, this reality is still objective, for I did not create them, I merely received them through my education. How many times it happens, moreover, that we are ignorant of the details of the obligations incumbent upon us, and that in order to get to know them we must consult the law and its authorised interpreters! Similarly, the church-member found the beliefs and practices of his religious life ready-made at birth; if they existed before him, they existed externally to him. The system of signs I use to express my thought, the system of currency I employ to pay my debts, the instruments of credit I utilise in my commercial relations, the practices followed in my profession, etc., function independently of whatever use I make of them. If one were to take, one after the other, all of the individuals who compose society, the preceding statement could be repeated of all of them. Here, then, are ways of acting, thinking, and feeling that present the remarkable property of existing outside the individual consciousness.

These types of conduct or thought are not only external to the individual but are, moreover, endowed with an imperative and coercive power, by virtue of which they impose themselves upon him, independent of his individual will. Of course, when I conform to them wholeheartedly, this coercion is felt only slightly, if at all, as it is unnecessary. But it is nevertheless an intrinsic characteristic of these facts; this is shown by the way in which it asserts itself as soon as I attempt to resist it. If I attempt to violate legal rules they react against me so as to prevent my act if possible, or to nullify my violation by restoring the damage, if it has been carried out but can be rectified; or to demand expiation if it cannot be compensated for otherwise.

Thus we are able to conceptualise in a precise way the field of sociology. It comprises only a limited group of phenomena. A social fact is to be recognised by the power of external coercion which it exercises or is capable of exercising over individuals; and the presence of this power can be recognised in turn either by the existence of some definite sanction or by the resistance offered against every individual act that tends to contravene it. . . .

The proposition according to which social facts are to be treated as things—which is the very foundation of our method—is one which has stimulated great opposition. It has been considered paradoxical and scandalous for us to assimilate the realities of the social world to those of the external world. Such criticism involves a singular misunderstanding of the meaning and application of this assimilation: the object of this was not to reduce the higher to the lower forms of being, but on the contrary to claim for the higher forms a degree of reality at least equal to that which is readily granted to the lower. We do not say that social facts are material things, but that they are things by the same right as material things, although they differ from them in type.

Just what is a "thing"? A thing differs from an idea in the same way as that which we know from without differs from that which we know from within. A thing is any object of knowledge which is not naturally controlled by the intellect, which cannot be adequately grasped by a simple process of mental activity. It can only be understood by the mind on condition that the mind goes outside itself by means of observations and experiments, which move progressively from the more external and immediately accessible characteristics to the less visible and more deep-lying. To treat the facts of a certain order as things thus is not to place them in a particular category of reality, but to assume a certain mental attitude toward them; it is to approach the study of them on the principle that we are absolutely ignorant of their nature, and that their characteristic properties, like the unknown causes on which they depend, cannot be discovered by even the most careful introspection.

With the terms thus defined, our proposition, far from being a paradox, could almost pass for a truism if it were not too often misunderstood in the human sciences and especially in sociology. Indeed, one might say in this sense that, with the possible exception of the case of mathematics, every object of science is a thing. In mathematics, since we proceed from simple to more complex concepts it is sufficient to depend upon mental processes which are purely internal in character. But in the case of "facts" properly so called, these are, at the moment when we undertake to study them scientifically, necessarily unknown *things* of which we are ignorant; for the representations which we have been able to make of them in the course of our life, having been made uncritically and unmethodically, are devoid of scientific value, and must be discarded. The facts of individual psychology themselves have this character and must be seen in this way. For although they are by definition purely mental, our consciousness of them reveals to us neither their real nature nor their genesis. It allows us to know them up to a certain point, just as our sensory knowledge gives us a certain familiarity with heat or light, sound or electricity; it gives us confused, fleeting, subjective impressions of them, but no clear and scientific notions or explanatory concepts. It is precisely for this reason that there has been founded in the course of this century an objective psychology whose fundamental purpose is to study mental facts from the outside, that is to say as things.

This is all the more necessary in the case of social facts, for consciousness is even more helpless in knowing them than in knowing its own life. It might be objected that since social facts are our own creations, we have only to look into our own mind in order to know what we put into them and how we formed them. But, in the first place, the greater part of our social institutions was bequeathed to us already formed by previous generations. We ourselves took no part in their formation, and consequently we cannot by introspection discover the causes which brought them into being. Furthermore, when we have in fact collaborated in their genesis, we can only with difficulty obtain even a very confused and a very distorted perception of the true nature of our action and the causes which determined it. When it is merely a matter of our private acts we know very imperfectly the relatively simple motives that guide us. We believe ourselves disinterested when we act egoistically; we think we are motivated by hate when we are yielding to love, that we obey reason when we are the slaves of irrational prejudices, etc. How, then, should we be able to discern with greater clarity the much more complex causes from which collective acts proceed? For, at the very least, each one of us participate in them only as an infinitesimal unit; a huge number of others collaborate with us, and what takes place in these other minds escapes us.

Thus our principle implies no metaphysical conception, no speculation about the fundamental nature of being. What it demands is that the sociologist put himself in the same state of mind as physicists, chemists, or physiologists, when they enquire into a hitherto unexplored region of the scientific domain. When he penetrates the social world, he must be aware that he is penetrating the unknown. He must feel himself in the presence of facts whose laws are as unsuspected as were those of life before the development of biology; he must be prepared for discoveries which will surprise and disconcert him. . . .

If we have said that societies are in nature, we have sought with no lesser determination to show that social nature is *sui generis,* that it is irreducible, not only to physical nature, but even to the psychic nature of the individual. To declare that societies are natural things, that collective events submit to necessary laws, is thus not to hold that there is nothing new or different in the world. No one has made greater efforts than us to show that the characteristic changes in social life are quite real and that, in a general way, the diversity of things is no mere appearance. In this, moreover, we do no more than to follow the path opened up by the founder of positive sociology, by Auguste Comte, who even went so far as to point to a radical discrepancy between the different realms of nature, and even between the different animal species.

Karl Marx

Karl Marx (1818–1883) has this in common with Freud—that they both claimed to have found a new science of human nature (social and individual, respectively) and that both were concerned not merely to explain human phenomena, but to apply their new understanding to liberate people from the social or psychological forces that prevented their free development. In Marx's famous words, the point was not just to interpret the world, but to change it.

These excerpts are from Marx's early works, from the *Economic and Philosophical Manuscripts,* the Introduction to *Towards a Critique of Hegel's Philosophy of Right* (both of 1844) and *The German Ideology* (written with Engels shortly afterwards). The translations are those reprinted in *Karl Marx: Selected Writings,* edited by David McLellan (Oxford University Press, 1977). Most of Marx's youthful writing has become known only in the last thirty years or so; his huge influence on the twentieth century has been exerted largely through the massive *Capital* of 1867 and other works such as the *Communist Manifesto* of 1848. It has sometimes been claimed (especially by the guardians of orthodoxy under the communist regimes) that these early works, in which the influence of the Hegelian style of philosophizing is more prominent, do not represent true Marxism. But scholarly opinion has come to recognize a continuity in Marx's work, finding the philosophical theme of alienation developed in the more "scientific" economic and sociological detail of the later works.

In the first passage reprinted here, we see Marx enthusiastically endorsing Feuerbach's way of debunking religion; it also represents his attitude to Hegel's philosophy as an expression of real human problems in terms of illusory abstractions. The other excerpts have been arranged into the pattern of theory, diagnosis, and prescription. It is apparent that Marx's "scientific" analysis of alienation as "a purely economic fact" involves a passionate condemnation of the capitalist system of his day.

DOWN FROM HEAVEN TO EARTH

As far as Germany is concerned, the criticism of religion is essentially complete, and the criticism of religion is the presupposition of all criticism.

The profane existence of error is compromised as soon as its heavenly *oratio pro aris et focis* [prayer for hearth and home] is refuted. Man has found in the imaginary reality of heaven where he looked for a superman only the reflection of his own self. He will therefore no longer be inclined to find only the appearance of himself, the non-man, where he seeks and must seek his true reality.

The foundation of irreligious criticism is this: man makes religion, religion does not make man. Religion is indeed the self-consciousness and self-awareness of man who either has not yet attained to himself or has already lost himself again. But man is no abstract being squatting outside the world. Man is the world of man, the state, society. This state, this society, produces religion's inverted attitude to the world, because they are an inverted world themselves. Religion is the general theory of this world, its encyclopaedic compendium, its logic in popular form, its spiritual *point d'honneur,* its enthusiasm, its moral sanction, its solemn complement, its universal basis for consolation and justification. It is the imaginary realization of the human essence, because the human essence possesses no true reality. Thus, the struggle against religion is indirectly the struggle against the world whose spiritual aroma is religion.

Religious suffering is at the same time an expression of real suffering and a protest against real suffering. Religion is the sigh of the oppressed creature, the feeling of a heartless world, and the soul of soulless circumstances. It is the opium of the people.

The abolition of religion as the illusory happiness of the people is the demand for their real happiness. The demand to give up the illusions about their condition is a demand to give up a condition that requires illusion. The criticism of religion is therefore the germ of the criticism of the valley of tears whose halo is religion.

Criticism has plucked the imaginary flowers from the chains not so that man may bear chains without any imagination or comfort, but so that he may throw away the chains and pluck living flowers. The criticism of religion disillusions man so that he may think, act, and fashion his own reality as a disillusioned man come to his senses; so that he may revolve around himself as his real sun. Religion is only the illusory sun which revolves around man as long as he does not revolve around himself.

It is therefore the task of history, now the truth is no longer in the beyond, to establish the truth of the here and now. The first task of philosophy, which is in the service of history, once the holy form of human self-alienation has been discovered, is to discover self-alienation in its unholy forms. The criticism of heaven is thus transformed into the criticism of earth, the criticism of religion into the criticism of law, and the criticism of theology into the criticism of politics. . . .

A NATURALISTIC APPROACH TO MAN

. . . [C]onsistent naturalism or humanism is distinguished from both idealism and materialism and constitutes at the same time their unifying truth. We see also how only naturalism is capable of understanding the process of world history.

Man is a directly natural being. As a living natural being he is on the one hand equipped with natural vital powers and is an active, natural being. These powers of his are dispositions, capacities, instincts. On the other hand, man as a natural, corporeal, sensuous, objective being is a passive, dependent, and limited being, like animals and plants, that is, the objects of his instincts are exterior to him and independent of him and yet they are objects of his need, essential objects that are indispensable for the exercise and confirmation of his faculties. The fact that man is an embodied, living, real, sentient objective being means that he has real, sensuous objects as the objects of his life-expression. In other words, he can only express his being in real, sensuous objects. To be objective, natural and sentient and to have one's object, nature and sense outside oneself or oneself to be object, nature and sense for a third person are identical. Hunger is a natural need; so it needs a natural object outside itself to satisfy and appease it. Hunger is the objective need of a body for an exterior object in order to be complete and express its being. The sun is the object of the plant, an indispensable object that confirms its life, just as the plant is the object of the sun in that it is the expression of the sun's life-giving power and objective faculties. . . .

But man is not only a natural being, he is a human natural being. This means that he is a being that exists for himself, thus a species-being that must confirm and exercise himself as such in his being and knowledge. Thus human objects are not natural objects as they immediately present themselves nor is human sense, in its purely objective existence, human sensitivity and human objectivity. Neither nature in its objective aspect nor in its subjective aspect is immediately adequate to the human being. And as everything natural must have an origin, so man too has his process of origin, history, which can, however, be known by him and thus is a conscious process of origin that transcends itself. History is the true natural history of man. . . .

The animal is immediately one with its vital activity. It is not distinct from it. They are identical. Man makes his vital activity itself into an object of his will and consciousness. He has a conscious vital activity. He is not immediately identical to any of his characterizations. Conscious vital activity differentiates man immediately from animal vital activity. It is this and this alone that makes man a species-being. He is only a conscious being, that is, his own life is an object to him, precisely because he is a species-being. This is the only reason for his activity being free activity. Alienated labour reverses the relationship so that, just because he is a conscious being, man makes his vital activity and essence a mere means to his existence.

The practical creation of an objective world, the working over of inorganic nature, is the confirmation of man as a conscious species-being, that is, as a being that relates to the species as to himself and to himself as to the species. It is true that the animal, too, produces. It builds itself a nest, a dwelling, like the bee, the beaver, the ant, etc. But it only produces what it needs immediately for itself or its offspring; it produces one-sidedly whereas man produces universally; it produces only under the pressure of immediate physical need, whereas man produces freely from physical need and only truly produces when he is thus free; it produces only itself whereas man reproduces the whole of nature. Its product belongs immediately to its physical body whereas man can freely separate himself from his product. The animal only fashions things according to the standards and needs of the species it belongs to, whereas man knows how to produce according to the measure of every species and knows everywhere how to apply its inherent standard to the object; thus man also fashions things according to the laws of beauty.

Thus it is in the working over of the objective world that man first really affirms himself as a species-being. This production is his active species-life. Through it nature appears as his work and his reality. The object of work is therefore the objectification of the species-life of man; for he duplicates himself not only intellectually, in his mind, but also actively in reality and thus can look at his image in a world he has created.

THE MATERIALIST CONCEPTION OF HISTORY

The premisses from which we begin are not arbitrary ones, not dogmas, but real premisses from which abstraction can only be made in the imagination. They are the real individuals, their activity and the material conditions under which they live, both those which they find already existing and those produced by their activity. These premisses can thus be verified in a purely empirical way.

The first premiss of all human history is, of course, the existence of living human individuals. Thus the first fact to be established is the physical organization of these individuals and their consequent relation to the rest of nature. Of course, we cannot here go either into the actual physical nature of man, or into the natural conditions in which man finds himself—geological, oro-hydrographical, climatic, and so on. The writing of history must always set out from these natural bases and their modification in the course of history through the action of men.

Men can be distinguished from animals by consciousness, by religion, or anything else you like. They themselves begin to distinguish themselves from animals as soon as they begin to produce their means of subsistence, a step which is conditioned by their physical organization. By producing their means of subsistence men are indirectly producing their actual material life.

The way in which men produce their means of subsistence depends first of all on the nature of the actual means of subsistence they find in existence and have to reproduce, This mode of production must not be considered simply as being the production of the physical existence of the individuals. Rather it is a definite form of activity of these individuals, a definite form of expressing their life, a definite mode of life on their part. As individuals express their life, so they are. What they are, therefore, coincides with their production, both with *what* they produce and with *how* they produce. The nature of individuals thus depends on the material conditions determining their production.

This production only makes its appearance with the increase of population. In its turn this presupposes the intercourse of individuals with one another. The form of this intercourse is again determined by production.

The relations of different nations among themselves depend upon the extent to which each has developed its productive forces, the division of labour, and internal intercourse. This statement is generally recognized. But not only the relation of one nation to others, but also the whole internal structure of the nation itself depends on the stage of development reached by its production and its internal and external intercourse. How far the productive forces of a nation are developed is shown most manifestly by the degree to which the division of labour has been carried. Each new productive force, in so far as it is not merely a quantitative extension of productive forces already known (for instance the bringing into cultivation of fresh land), causes a further development of the division of labour.

The division of labour inside a nation leads at first to the separation of industrial and commercial from agricultural labour, and hence to the separation of town and country and to the conflict of their interests. Its further development leads to the separation of commercial from industrial labour. At the same time, through the division of labour inside these various branches there develop various divisions among the individuals co-operating in definite kinds of labour. The relative position of these individual groups is determined by the methods employed in agriculture, industry, and commerce (patriarchalism, slavery, estates, classes). These same conditions are to be seen (given a more developed intercourse) in the relations of different nations to one another.

The various stages of development in the division of labour are just so many different forms of ownership, i.e., the existing stage in the division of labour determines also the relations of individuals to one another with reference to the material, instrument, and product of labour.

The first form of ownership is tribal ownership. It corresponds to the undeveloped stage of production, at which a people lives by hunting and fishing, by the rearing of beasts, or, in the highest stage, agriculture. In the latter case it presupposes a great mass of uncultivated stretches of land. The division of labour is at this stage still very elementary and is confined to a further extension of the natural division of labour existing in the family. The social structure is, therefore, limited to an extension of the family; patriarchal family chieftains, below them the members of the tribe, finally slaves. The slavery latent in the family only develops gradually with the increase of population, the growth of wants, and with the extension of external relations, both of war and of barter.

The second form is the ancient communal and State ownership which proceeds especially from the union of several tribes into a city by agreement or by conquest, and which is still accompanied by slavery. Beside communal ownership we already find movable, and later also immovable, private property developing, but as an abnormal form subordinate to communal ownership. The citizens hold power over their labouring slaves only in their community, and on this account alone, therefore, they are bound to the form of communal ownership. It is the communal private property which compels the active citizens to remain in this spontaneously derived form of association over against their slaves. For this reason the whole structure of society based on this communal ownership, and with it the power of the people, decays in the same measure as, in particular, immovable private property evolves. The division of labour is already more developed. We already find the antagonism of town and country; later the antagonism between those states which represent town interests and those which represent country interests, and inside the towns themselves the antagonisms between industry and maritime commerce. The class relation between citizens and slaves is now completely developed.

With the development of private property, we find here for the first time the same conditions which we shall find again, only on a more extensive scale, with modern private property. On the one hand, the concentration of private property, which began very early in Rome (as the Licinian agrarian law proves) and proceeded very rapidly from the time of the civil wars and especially under the Emperors; on the other hand, coupled with this, the transformation of the plebeian small peasantry into a proletariat, which, however, owing to its intermediate position between propertied citizens and slaves, never achieved an independent development.

The third form of ownership is feudal or estate property. If antiquity started out from the town and its little territory, the Middle Ages started out from the country. This differing starting-point was determined by the sparseness of the population at that time, which was scattered over a large area and which received no large increase from the conquerors. In contrast to Greece and Rome, feudal development at the outset, therefore, extends over a much wider territory, prepared by the Roman conquests and the spread of agriculture at first associated with it. The last centuries of the declining Roman Empire and its conquest by the barbarians destroyed a number of productive forces; agriculture had declined, industry had decayed for want of a market, trade had died out or been violently suspended, the rural and urban population had decreased. From these conditions and the mode of organization of the conquest determined by them, feudal property developed under the influence of the Germanic military constitution. Like tribal and communal ownership, it is based again on a community; but the directly producing class standing over against it is not, as in the case of the ancient community, the slaves, but the enserfed small peasantry. As soon as feudalism is fully developed, there also arises antagonism towards the towns. The hierarchical structure of land-ownership, and the armed bodies of retainers associated with it, gave the nobility power over the serfs. This feudal organization was, just as much as the ancient communal ownership, an association against a subjected producing class; but the form of association and the relation to the direct producers were different because of the different conditions of production.

This feudal system of landownership had its counterpart in the towns in the shape of corporative property, the feudal organization of trades. Here property consisted chiefly in the labour of each individual person. The necessity for association against the organized robber barons, the need for communal covered markets in an age when the industrialist was at the same time a merchant, the growing competition of the escaped serfs swarming into the rising towns, the feudal structure of the whole country: these combined to bring about the guilds. The gradually accumulated small capital of individual craftsmen and their stable numbers, as against the growing population, evolved the relation of journeyman and apprentice, which brought into being in the towns a hierarchy similar to that in the country.

Thus the chief form of property during the feudal epoch consisted on the one hand of landed property with serf labour chained to it, and on the other of the labour of the individual with small capital commanding the labour of journeymen. The organization of both was determined by the restricted conditions of production— the small-scale and primitive cultivation of the land and the craft type of industry. There was little division of labour in the heyday of feudalism. Each country bore in itself the antithesis of town and country; the division into estates was certainly strongly marked; but apart from the differentiation of princes, nobility, clergy, and peasants in the country, and masters, journeymen, apprentices, and soon also the rabble of casual labourers in the towns, no division of importance took place. In agriculture it was rendered difficult by the strip-system, beside which the cottage industry of the peasants themselves emerged. In industry there was no division of labour at all in the individual trades themselves, and very little between them. The separation of industry and commerce was found already in existence in older towns; in the newer it only developed later, when the towns entered into mutual relations.

The grouping of larger territories into feudal kingdoms was a necessity for the landed nobility as for the towns. The organization of the ruling class, the nobility, had, therefore, everywhere a monarch at its head.

The fact is, therefore, that definite individuals who are productively active in a definite way enter into these definite social and political relations. Empirical observation must in each separate instance bring out empirically, and without any mystification and speculation, the connection of the social and political structure with production. The social structure and the State are continually evolving out of the life-process of definite individuals, but of individuals, not as they may appear in their own or other people's imagination, but as they really are, i.e. as they operate, produce materially, and hence as they work under definite material limits, presuppositions, and conditions independent of their will.

The production of ideas, of conceptions, of consciousness, is at first directly interwoven with the material activity and the material intercourse of men, the language of real life. Conceiving, thinking, the mental intercourse of men, appear at this stage as the direct efflux of their material behaviour. The same applies to mental production as expressed in the language of politics, laws, morality, religion, metaphysics, etc., of a people. Men are the producers of their conceptions, ideas, etc.—real, active men, as they are conditioned by a definite development of their productive forces and of the intercourse corresponding to these, up to its furthest forms. Consciousness can never be anything else than conscious existence, and the existence of men is their actual life-process. If in all ideology men and their circumstances appear upside-down as in a *camera obscura,* this phenomenon arises just as much from their historical life-process as the inversion of objects on the retina does from their physical life-process.

In direct contrast to German philosophy which descends from heaven to earth, here we ascend from earth to heaven. That is to say, we do not set out from what men say, imagine, conceive, nor from men as narrated, thought of, imagined, conceived, in order to arrive at men in the flesh. We set out from real, active men, and on the basis of their real life-process we demonstrate the development of the ideological reflexes and echoes of this life-process. The phantoms formed in the human brain are also, necessarily, sublimates of their material life-process, which is empirically verifiable and bound to material premisses. Morality, religion, metaphysics, all the rest of ideology and their corresponding forms of consciousness, thus no longer retain the semblance of independence. They have no history, no development; but men, developing their material production and their material intercourse, alter, along with this their real existence, their thinking and the products of their thinking. Life is not determined by consciousness, but consciousness by life. In the first method of approach the starting-point is consciousness taken as the living individual; in the second method, which conforms to real life, it is the real living individuals themselves, and consciousness is considered solely as their consciousness.

This method of approach is not devoid of premises. It starts out from the real premises and does not abandon them for a moment. Its premises are men, not in any fantastic isolation and rigidity, but in their actual, empirically perceptible process of development under definite conditions. As soon as this active life-process is described, history ceases to be a collection of dead facts as it is with the empiricists (themselves still abstract), or an imagined activity of imagined subjects, as with the idealists.

Where speculation ends—in real life—there real, positive science begins: the representation of the practical activity, of the practical process of development of men. Empty talk about consciousness ceases, and real knowledge has to take its place. When reality is depicted, philosophy as an independent branch of knowledge loses its medium of existence. At the best its place can only be taken by a summing-up of the most general results, abstractions which arise from the observation of the historical development of men. Viewed apart from real history, these abstractions have in themselves no value whatsoever. They can only serve to facilitate the arrangement of historical material, to indicate the sequence of its separate strata. But they by no means afford a recipe or schema, as does philosophy, for neatly trimming the epochs of history. On the contrary, our difficulties begin only when we set about the observation and the arrangement—the real depiction—of our historical material, whether of a past epoch or of the present. . . .

ALIENATION IN CAPITALIST SOCIETY

We started from the presuppositions of political economy. We accepted its vocabulary and its laws. We presupposed private property, the separation of labour, capital, and land, and likewise of wages, profit, and ground rent; also division of labour; competition; the concept of exchange value, etc. Using the very words of political economy we have demonstrated that the worker is degraded to the most miserable sort of commodity; that the misery of the worker is in inverse proportion to the power and size of his production; that the necessary result of competition is the accumulation of capital in a few hands, and thus a more terrible restoration of monopoly; and that finally the distinction between capitalist and landlord, and that between peasant and industrial worker disappears and the whole of society must fall apart into the two classes of the property owners and the propertyless workers.

Political economy starts with the fact of private property, it does not explain it to us. It conceives of the material process that private property goes through in reality in general abstract formulas which them have for it a value of laws. It does not understand these laws, i.e. it does not demonstrate how they arise from the nature of private property. Political economy does not afford us any explanation of the reason for the separation of labour and capital, of capital and land. When, for example, political economy defines the relationship of wages to profit from capital, the interest of the capitalist is the ultimate court of appeal, that is, it presupposes what should be its result. In the same way competition enters the argument everywhere. It is explained by exterior circumstances. But political economy tells us nothing about how far these exterior, apparently fortuitous circumstances are merely the expression of a necessary development. We have seen how it regards exchange itself as something fortuitous. The only wheels that political economy sets in motion are greed and war among the greedy, competition.

It is just because political economy has not grasped the connections in the movement that new contradictions have arisen in its doctrines, for example, between that of monopoly and that of competition, freedom of craft and corporations, divisions of landed property and large estates. For competition, free trade, and the division of landed property were only seen as fortuitous circumstances created by will and force, not developed and comprehended as necessary, inevitable, and natural results of monopoly, corporations, and feudal property.

So what we have to understand now is the essential connection of private property, selfishness, the separation of labour, capital, and landed property, of exchange and competition, of the value and degradation of man, of monopoly and competition, etc.—the connection of all this alienation with the money system.

Let us not be like the political economist who, when he wishes to explain something, puts himself in an imaginary original state of affairs. Such an original stage of affairs explains nothing. He simply pushes the question back into a grey and nebulous distance. He presupposes as a fact and an event what he ought to be deducing, namely, the necessary connection between the two things, for example, between the division of labour and exchange. Similarly, the theologian explains the origin of evil through the fall, i.e. he presupposes as an historical fact what he should be explaining.

We start with a contemporary fact of political economy:

The worker becomes poorer the richer is his production, the more it increases in power and scope. The worker becomes a commodity that is all the cheaper the more commodities he creates. The depreciation of the human world progresses in direct proportion to the increase in value of the world of things. Labour does not only produce commodities; it produces itself and the labourer as a commodity and that to the extent to which it produces commodities in general.

What this fact expresses is merely this: the object that labour produces, its product, confronts it as an alien being, as a power independent of the producer. The product of labour is labour that has solidified itself into an object, made itself into a thing, the objectification of labour. The realization of labour is its objectification. In political economy this realization of labour appears as a loss of reality for the worker, objectification as a loss of the object or slavery to it, and appropriation as alienation, as externalization.

The realization of labour appears as a loss of reality to an extent that the worker loses his reality by dying of starvation. Objectification appears as a loss of the object to such an extent that the worker is robbed not only of the objects necessary for his life but also of the objects of his work. Indeed, labour itself becomes an object he can only have in his power with the greatest of efforts and at irregular intervals. The appropriation of the object appears as alienation to such an extent that the more objects the worker produces, the less he can possess and the more he falls under the domination of his product, capital.

All these consequences follow from the fact that the worker relates to the product of his labour as to an alien object. For it is evident from this presupposition that the more the worker externalizes himself in his work, the more powerful becomes the alien, objective world that he creates opposite himself, the poorer he becomes himself in his inner life and the less he can call his own. It is just the same in religion. The more man puts into God, the less he retains in himself. The worker puts his life into the object and this means that it no longer belongs to him but to the object. So the greater this activity, the more the worker is without an object. What the product of his labour is, that he is not. So the greater this product the less he is himself. The externalization of the worker in his product implies not only that his labour becomes an object, an exterior existence but also that it exists outside him, independent and alien, and becomes a self-sufficient power opposite him, that the life that he has lent to the object affronts him, hostile and alien.

Let us now deal in more detail with objectification, the production of the worker, and the alienation, the loss of the object, his product, which is involved in it.

The worker can create nothing without nature, the sensuous exterior world. It is the matter in which his labour realizes itself, in which it is active, out of which and through which it produces.

But as nature affords the means of life for labour in the sense that labour cannot live without objects on which it exercises itself, so it affords a means of life in the narrower sense, namely the means for the physical subsistence of the worker himself.

Thus the more the worker appropriates the exterior world of sensuous nature by his labour, the more he doubly deprives himself of the means of subsistence, firstly since the exterior sensuous world increasingly ceases to be an object belonging to his work, a means of subsistence for his labour; secondly, since it increasingly ceases to be a means of subsistence in the direct sense, a means for the physical subsistence of the worker.

Thus in these two ways the worker becomes a slave to his object: firstly he receives an object of labour, that is he receives labour, and secondly, he receives the means of subsistence. Thus it is his object that permits him to exist first as a worker and secondly as a physical subject. The climax of this slavery is that only as a worker can he maintain himself as a physical subject and it is only as a physical subject that he is a worker.

(According to the laws of political economy the alienation of the worker in his object is expressed as follows: the more the worker produces the less he has to consume, the more values he creates the more valueless and worthless he becomes, the more formed the product the more deformed the worker, the more civilized the product, the more barbaric the worker, the more powerful the work the more powerless becomes the worker, the more cultured the work the more philistine the worker becomes and more of a slave to nature.)

Political economy hides the alienation in the essence of labour by not considering the immediate relationship between the worker (labour) and production. Labour produces works of wonder for the rich, but nakedness for the worker. It produces palaces, but only hovels for the worker; it produces beauty, but cripples the worker; it replaces labour by machines but throws a part of the workers back to a barbaric labour and turns the other part into machines. It produces culture, but also imbecility and cretinism for the worker.

The immediate relationship of labour to its products is the relationship of the worker to the objects of his production. The relationship of the man of means to the objects of production and to production itself is only a consequence of this first relationship. And it confirms it. We shall examine this other aspect later.

So when we ask the question: what relationship is essential to labour, we are asking about the relationship of the worker to production.

Up to now we have considered only one aspect of the alienation or externalization of the worker, his relationship to the products of his labour. But alienation shows itself not only in the result, but also in the act of production, inside productive activity itself. How would the worker be able to affront the product of his work as an alien being if he did not alienate himself in the act of production itself? For the product is merely the summary of the activity of production. So if the product of labour is externalization, production itself must be active externalization, the externalization of activity, the activity of externalization. The alienation of the object of labour is only the résumé of the alienation, the externalization in the activity of labour itself.

What does the externalization of labour consist of then?

Firstly, that labour is exterior to the worker, that is, it does not belong to his essence. Therefore he does not confirm himself in his work, he denies himself, feels miserable instead of happy, deploys no free physical and intellectual energy, but mortifies his body and ruins his mind. Thus the worker only feels a stranger. He is at home when he is not working and when he works he is not at home. His labour is therefore not voluntary but compulsory, forced labour. It is therefore not the satisfaction of a need but only a means to satisfy needs outside itself. How alien it really is is very evident from the fact that when there is no physical or other compulsion, labour is avoided like the plague. External labour, labour in which man externalizes himself, is a labour of self-sacrifice and mortification. Finally, the external character of labour for the worker shows itself in the fact that it is not his own but someone else's, that it does not belong to him, that he does not belong to himself in his labour but to someone else. As in religion the human imagination's own activity, the activity of man's head and his heart, reacts independently on the individual as an alien activity of gods or devils, so the activity of the worker is not his own spontaneous activity. It belongs to another and is the loss of himself.

The result we arrive at then is that man (the worker) only feels himself freely active in his animal functions of eating, drinking, and procreating, at most also in his dwelling and dress, and feels himself an animal in his human functions.

Eating, drinking, procreating, etc., are indeed truly human functions. But in the abstraction that separates them from the other round of human activity and makes them into final and exclusive ends they become animal.

We have treated the act of alienation of practical human activity, labour, from two aspects. (1) The relationship of the worker to the product of his labour as an alien object that has power over him. This relationship is at the same time the relationship to the sensuous exterior world and to natural objects as to an alien and hostile world opposed to him. (2) The relationship of labour to the act of production inside labour. This relationship is the relationship of the worker to his own activity as something that is alien and does not belong to him; it is activity that is passivity, power that is weakness, procreation that is castration, the worker's own physical and intellectual energy, his personal life (for what is life except activity?) as an activity directed against himself, independent of him and not belonging to him. It is self-alienation, as above it was the alienation of the object.

We now have to draw a third characteristic of alienated labour from the two previous ones.

Man is a species-being not only in that practically and theoretically he makes both his own and other species into his objects, but also, and this is only another way of putting the same thing, he relates to himself as to the present, living species, in that he relates to himself as to a universal and therefore free being.

Both with man and with animals the species-life consists physically in the fact that man (like animals) lives from inorganic nature, and the more universal man is than animals the more more universal is the area of inorganic nature from which he lives. From the theoretical point of view, plants, animals, stones, air, light, etc., form part of human consciousness, partly as objects of natural science, partly as objects of art; they are his intellectual inorganic nature, his intellectual means of subsistence, which he must first prepare before he can enjoy and assimilate them. From the practical point of view, too, they form a part of human life and activity. Physically man lives solely from these products of nature, whether they appear as food, heating, clothing, habitation, etc. The universality of man appears in practice precisely in the universality that makes the whole of nature into his inorganic body in that it is both (i) his immediate means of subsistence and also (ii) the material object and tool of his vital activity. Nature is the inorganic body of a man, that is, in so far as it is not itself a human body. That man lives from nature means that nature is his body with which he must maintain a constant interchange so as not to die. That man's physical and intellectual life depends on nature merely means that nature depends on itself, for man is a part of nature.

While alienated labour alienates (1) nature from man, and (2) man from himself, his own active function, his vital activity, it also alienates the species from man; it turns his species-life into a means towards his individual life. Firstly it alienates species-life and individual life, and secondly in its abstraction it makes the latter into the aim of the former which is also conceived of in its abstract and alien form. For firstly, work, vital activity, and productive life itself appear to man only as a means to the satisfaction of a need, the need to preserve his physical existence. But productive life is species-life. It is life producing life. The whole character of a species, its generic character, is contained in its manner of vital activity,and free conscious activity is the species-characteristic of man. Life itself appears merely as a means to life. . . .

. . . Therefore when alienated labour tears from man the object of his production, it also tears from him his species-life, the real objectivity of his species and turns the advantage he has over animals into a disadvantage in that his inorganic body, nature, is torn from him.

Similarly, in that alienated labour degrades man's own free activity to a means, it turns the species-life of man into a means for his physical existence.

Thus consciousness, which man derives from his species, changes itself through alienation so that species-life becomes a means for him.

Therefore alienated labour:

(3) Makes the species-being of man, both nature and the intellectual faculties of his species, into a being that is alien to him, into a means for his individual existence. It alienates from man his own body, nature exterior to him, and his intellectual being, his human essence.

(4) An immediate consequence of man's alienation from the product of his work, his vital activity and his species-being, is the alienation of man from man. When man is opposed to himself, it is another man that is opposed to him. What is valid for the relationship of a man to his work, of the product of his work and himself, is also valid for the relationship of man to other men and of their labour and the objects of their labour.

In general, the statement that man is alienated from his species-being, means that one man is alienated from another as each of them is alienated from the human essence.

The alienation of man and in general of every relationship in which man stands to himself is first realized and expressed in the relationship with which man stands to other men.

Thus in the situation of alienated labour each man measures his relationship to other men by the relationship in which he finds himself placed as a worker.

We began with a fact of political economy, the alienation of the worker and his production. We have expressed this fact in conceptual terms: alienated, externalized labour. We have analysed this concept and thus analysed a purely economic fact. . . .

Up till now we have considered the relationship only from the side of the worker and we will later consider it from the side of the non-worker.

Thus through alienated, externalized labour the worker creates the relationship to this labour of a man who is alien to it and remains exterior to it. The relationship of the worker to his labour creates the relationship to it of the capitalist, or whatever else one wishes to call the master of the labour. Private property is thus the product, result, and necessary consequence of externalized labour, of the exterior relationship of the worker to nature and to himself.

Thus private property is the result of the analysis of the concept of externalized labour, i.e. externalized man, alienated work, alienated life, alienated man.

We have, of course, obtained the concept of externalized labour (externalized life) from political economy as the result of the movement of private property. But it is evident from the analysis of this concept that, although private property appears to be the ground and reason for externalized labour, it is rather a consequence of it, just as the gods are originally not the cause but the effect of the aberration of the human mind, although later this relationship reverses itself.

It is only in the final culmination of the development of private property that these hidden characteristics come once more to the fore, in that firstly it is the product of externalized labour and secondly it is the means through which labour externalizes itself, the realization of this externalization. . . .

COMMUNISM AS THE OVERCOMING OF ALIENATION

. . . We can easily see how necessary it is that the whole revolutionary movement should find not so much its empirical as its theoretical basis in the development of private property and particularly the economic system.

This material, immediately sensuous, private property, is the material, sensuous expression of man's alienated life. Its movement of production and consumption is the sensuous revelation of the movement of all previous production, i.e. the realization or reality of man. Religion, family, state, law, morality, science, and art are only particular forms of production and fall under its general law. The positive abolition of private property and the appropriation of human life is therefore the positive abolition of all alienation, thus the return of man out of religion, family, state, etc., into his human, i.e. social being. Religious alienation as such occurs only in man's interior consciousness, but economic alienation is that of real life, and its abolition therefore covers both aspects. It is obvious that the movement begins differently with different peoples according to whether the actual conscious life of the people is lived in their minds or in the outer world, is an ideal or a real life. Communism begins immediately with atheism (Owen) and atheism is at first still very far from being communism, for this atheism is still rather an abstraction. The philanthropy of atheism is therefore at first only an abstract philosophical philanthropy whereas that of communism is immediately real and directly oriented towards action. . . .

. . . [T]here is communism as the positive abolition of private property and thus of human self-alienation and therefore the real reappropriation of the human essence by and for man. This is communism as the complete and conscious return of man conserving all the riches of previous development for man himself as a social, i.e. human being. Communism as completed naturalism is humanism and as completed humanism is naturalism. It is the genuine solution of the antagonism between man and nature and between man and man. It is the true solution of the struggle between existence and essence, between objectification and self-affirmation, between freedom and necessity, between individual and species. It is the solution to the riddle of history and knows itself to be this solution. . . .

. . . We have seen how, presupposing the positive supercession of private property, man produces man, himself and other men; how also the object, which is the direct result of his personal activity, is at the same time his own existence for other men and their existence for him. In the same way, both the materials of his labour and man its author are the result and at the same time the origin of the movement. (And private property is historically necessary precisely because there must be this origin.) So the general character of the whole movement is a social one; as society produces man as man, so it is produced by man. Activity and enjoyment are social both in their content and in their mode of existence; they are social activity and social enjoyment. The human significance of nature is only available to social man; for only to social man is nature available as a bond with other men, as the basis of his own existence for others and theirs for him, and as the vital element in human reality; only to social man is nature the foundation of his own human existence. Only as such has his natural existence become a human existence and nature itself become human. Thus society completes the essential unity of man and nature, it is the genuine resurrection of nature, the accomplished naturalism of man and the accomplished humanism of nature.

Social activity and social enjoyment by no means exist only in the form of a directly communal activity and directly communal enjoyment. But communal activity and enjoyment, i.e. activity and enjoyment that is expressed and confirmed in the real society of other men, will occur everywhere where this direct expression of sociability arises from the content of the activity or enjoyment and corresponds to its nature.

But even if my activity is a scientific one, etc., an activity that I can seldom perform directly in company with other men, I am still acting socially since I am acting as a man. Not only the material of my activity—like language itself for the thinker—is given to me as a social product, my own existence is social activity; therefore what I individually produce, I produce individually for society, conscious of myself as a social being.

My universal consciousness is only the theoretical form whose living form is the real community, society, whereas at the present time universal consciousness is an abstraction from real life and as such turns into its enemy. Thus the activity of my universal consciousness is as such my theoretical existence as a social being.

It is above all necessary to avoid restoring society as a fixed abstraction opposed to the individual. The individual is the social being. Therefore, even when the manifestation of his life does not take the form of a communal manifestation performed in the company of other men, it is still a manifestation and confirmation of social life. The individual and the species-life of man are not different, although, necessarily, the mode of existence of individual life is a more particular or a more general mode of species-life or the species-life is a more particular or more general individual life.

Man confirms his real social life in his species-consciousness and in his thought he merely repeats his real existence just as conversely his species-being is confirmed in his species-consciousness and exists for itself in its universality as a thinking being.

However much he is a particular individual (and it is precisely his particularity that makes him an individual and a truly individual communal being) man is just as much the totality, the ideal totality, the subjective existence of society as something thought and felt. Man exists also in reality both as the contemplation and true enjoyment of social existence and as the totality of human manifestations of life. . . .

THE NEED FOR REVOLUTION

Our investigation hitherto started from the instruments of production, and it has already shown that private property was a necessity for certain industrial stages. In *industrie extractive* [raw materials industry] private property still coincides with labour; in small industry and all agriculture up till now property is the necessary consequence of the existing instruments of production; in big industry the contradiction between the instrument of production and private property appears for the first time and is the product of big industry; moreover, big industry must be highly developed to produce this contradiction. And thus only with big industry does the abolition of private property become possible.

In big industry and competition the whole mass of conditions of existence, limitations, biases of individuals, are fused together into the two simplest forms: private property and labour. With money every form of intercourse, and intercourse itself, is considered fortuitous for the individuals. Thus money implies that all previous intercourse was only intercourse of individuals under particular conditions, not of individuals as individuals. These conditions are reduced to two: accumulated labour or private property, and actual labour. If both or one of these ceases, then intercourse comes to a standstill. The modern economists themselves, e.g. Sismondi, Cherbuliez, etc., oppose "association of individuals" to "association of capital." On the other hand, the individuals themselves are entirely subordinated to the division of labour and hence are brought into the most complete dependence on one another. Private property, in so far as within labour itself it is opposed to labour, evolves out of the necessity of accumulation, and has still, to begin with, rather the form of the communality; but in its further development it approaches more and more the modern form of private property. The division of labour implies from the outset the division of the conditions of labour, of tools and materials, and thus the splitting-up of accumulated capital among different owners, and thus, also, the division between capital and labour, and the different forms of property itself. The more the division of labour develops and accumulation grows, the sharper are the forms that this process of differentiation assumes. Labour itself can only exist on the premiss of this fragmentation.

Thus two facts are here revealed. First the productive forces appear as a world for themselves, quite independent of and divorced from the individuals, alongside the individuals: the reason for this is that the individuals, whose forces they are, exist split up and in opposition to one another, while, on the other hand, these forces are the only real forces in the intercourse and association of these individuals. Thus, on one hand, we have a totality of productive forces, which have, as it were, taken on a material form and are for the individuals no longer the forces of the individuals but of private property, and hence of the individuals only in so far as they are owners of private property themselves. Never, in any earlier period, have the productive forces taken on a form so indifferent to the intercourse of individuals as individuals, because their intercourse itself was formerly a restricted one. On the other hand, standing over against these productive forces, we have the majority of the individuals from whom these forces have been wrested away, and who, robbed thus of all real life-content, have become abstract individuals, but who are, however, only by this fact put into a position to enter into relation with one another as individuals.

The only connection which still links them with the productive forces and with their own existence—labour—has lost all semblance of self-activity and only sustains their life by stunting it. While in the earlier periods self-activity and the production of material life were separated, in that they devolved on different persons, and while, on account of the narrowness of the individuals themselves, the production of material life was considered as a subordinate mode of self-activity, they now diverge to such an extent that altogether material life appears at the end, and what produces this material life, labour (which is now the only possible but, as we see, negative form of self-activity), as the means.

Thus things have now come to such a pass that the individuals must appropriate the existing totality of productive forces, not only to achieve self-activity, but also merely to safeguard their very existence. This appropriation is first determined by the object to be appropriated, the productive forces, which have been developed to a totality and which only exist within a universal intercourse. From this aspect alone, therefore, this appropriation must have a universal character corresponding to the productive forces and the intercourse.

The appropriation of these forces is itself nothing more than the development of the individual capacities corresponding to the material instruments of production. The appropriation of a totality of instruments of production is, for this very reason, the development of a totality of capacities in the individuals themselves.

This appropriation is further determined by the persons appropriating. Only the proletarians of the present day, who are completely shut off from all self-activity, are in a position to achieve a complete and no longer restricted self-activity, which consists in the appropriation of a totality of productive forces and in the thus postulated development of a totality of capacities. All earlier revolutionary appropriations were restricted; individuals, whose self-activity was restricted by a crude instrument of production and a limited intercourse, appropriated this crude instrument of production, and hence merely achieved a new state of limitation. Their instrument of production became their property, but they themselves remained subordinate to the division of labour and their own instrument of production. In all expropriations up to now, a mass of individuals remained subservient to a single instrument of production; in the appropriation by the proletarians, a mass of instruments of production must be made subject to each individual, and property to all. Modern universal intercourse can be controlled by individuals, therefore, only when controlled by all.

This appropriation is further determined by the manner in which it must be effected. It can only be effected through a union, which by the character of the proletariat itself can again only be a universal one, and through a revolution, in which, on the one hand, the power of the earlier mode of production and intercourse and social organization is overthrown, and, on the other hand, there develops the universal character and the energy of the proletariat, without which the revolution cannot be accomplished; and in which, further, the proletariat rids itself of everything that still clings to it from its previous position in society.

Only at this stage does self-activity coincide with material life, which corresponds to the development of individuals into complete individuals and the casting-off of all natural limitations. The transformation of labour into self-activity corresponds to the transformation of the earlier limited intercourse into the intercourse of individuals as such. With the appropriation of the total productive forces through united individuals, private property comes to an end. While previously in history a particular condition always appeared as accidental, now the isolation of individuals and the particular private gain of each man have themselves become accidental. . . .

Finally, from the conception of history we have sketched we obtain these further conclusions: (1) In the development of productive forces there comes a stage when productive forces and means of intercourse are brought into being, which, under the existing relationships, only cause mischief, and are no longer productive but destructive forces (machinery and money); and connected with this a class is called forth, which has to bear all the burdens of society without enjoying its advantages, which, ousted from society, is forced into the most decided antagonism to all other classes; a class which forms the majority of all members of society, and from which emanates the consciousness of the necessity of a fundamental revolution, the communist consciousness, which may, of course, arise among the other classes too through the contemplation of the situation of this class. (2) The conditions under which definite productive forces can be applied are the conditions of the rule of a definite class of society, whose social power, deriving from its property, has its practical-idealistic expression in each case in the form of the State; and, therefore, every revolutionary struggle is directed against a class, which till then has been in power. (3) In all revolutions up till now the mode of activity always remained unscathed and it was only a question of a different distribution of this activity, a new distribution of labour to other persons, while the communist revolution is directed against the preceding mode of activity, does away with labour, and abolishes the rule of all classes with the classes themselves, because it is carried through by the class which no longer counts as a class in society, is not recognized as a class, and is in itself the expression of the dissolution of all classes, nationalities, etc. within present society; and (4) Both for the production on a mass scale of this communist consciousness, and for the success of the cause itself, the alteration of men on a mass scale is necessary, an alteration which can only take place in a practical movement, a revolution; this revolution is necessary, therefore, not only because the ruling class cannot be overthrown in any other way, but also because the class overthrowing it can only in a revolution succeed in ridding itself of all the muck of ages and become fitted to found society anew. . . .

. . . So where is the real possibility of a German emancipation?

We answer: in the formation of a class with radical chains, a class in civil society that is not a class of civil society, of a social group that is the dissolution of all social groups, of a sphere that has a universal character because of its universal sufferings and lays claim to no particular right, because it is the object of no particular injustice but of injustice in general. This class can no longer lay claim to a historical status, but only to a human one. It is not in a one-sided opposition to the consequences of the German political regime, it is in total opposition to its presuppositions. It is, finally, a sphere that cannot emancipate itself without emancipating itself from all other spheres of society and thereby emancipating these other spheres themselves. In a word, it is the complete loss of humanity and thus can only recover itself by a complete redemption of humanity. This dissolution of society, as a particular class, is the proletariat.

The proletariat is only beginning to exist in Germany through the invasion of the industrial movement. For it is not formed by the poverty produced by natural laws but by artificially induced poverty. It is not made up of the human masses mechanically oppressed by the weight of society, but of those who have their origin in society's brutal dissolution and principally the dissolution of the middle class, although, quite naturally, its ranks are gradually swelled by natural poverty and Germano-Christian serfdom.

When the proletariat proclaims the dissolution of the hitherto existing world order, it merely declares the secret of its own existence, since it is in fact the dissolution of this order. When it demands the negation of private property it is only laying down as a principle for society what society has laid down as a principle for the proletariat, what has already been incorporated in itself without its consent as the negative result of society. The proletarian thus finds that he has in relation to the world of the future the same right as the German king in relation to the world of the past, when he calls the people *his* people as he might call a horse *his* horse. When the king declares the people to be his private property he is only confirming that the private property owner is king.

As philosophy finds in the proletariat its material weapons, so the proletariat finds in philosophy its intellectual weapons, and as soon as the lightning of thought has struck deep into the virgin soil of the people, the emancipation of the Germans into men will be completed.

Let us summarize our results:

The only liberation of Germany that is practically possible is the liberation from the theoretical standpoint that declares man to be the highest being for man. In Germany emancipation from the Middle Ages is only possible as an emancipation from the partial overcoming of the Middle Ages. In Germany no form of slavery can be broken without every form of slavery being broken. Germany is thorough and cannot make a revolution without its being a thorough one. The emancipation of Germany is the emancipation of man. The head of this emancipation is philosophy, its heart is the proletariat. Philosophy cannot realize itself without transcending the proletariat, the proletariat cannot transcend itself without realizing philosophy.

Sigmund Freud

As he says in the first paragraph here, Freud (1856–1939) developed his new theory and treatment (both of which bear the name *psychoanalysis*) over a long period, in which his ideas changed considerably. Even after the fairly late writing reprinted here, there were further theoretical changes (notably, the introduction of the death instinct). But there is undoubtedly a strong continuity of basic themes running through all Freud's psychoanalytic work.

In his later period, Freud preferred to introduce his theory of the mind by the tripartite structure of *id, ego,* and *superego* (the latter is here kept in reserve until he comes to describe treatment in more detail). Earlier, he made a different distinction between conscious, preconscious, and unconscious elements in the mind. Earlier still, in the *Scientific Project* of 1895, he had tried to relate his embryonic psychological theories to postulated electrical processes in the brain, but, as he says here, he came to regard the question of the material embodiment of the mental functions as not of psychological interest.

Themes from Freud were criticized by B. F. Skinner and J. P. Sartre later in our century (from very different points of view), and discussed more sympathetically by S. Hampshire, as we shall see in the readings that follow.

These extracts are from "The Question of Lay-Analysis," first published in 1926 (translated in Volume XX of *The Standard Edition of the Complete Psychological Works of Sigmund Freud,* under the general editorship of James Strachey, Hogarth Press, 1953–1974; *The Question of Lay Analysis* is published in the United States by W. W. Norton). In this work, Freud defends the right to practice of psychoanalysts who were not medically qualified; the result is a stylish dialogue, in which he gives one of the clearest popular expositions of the main points of psychoanalytic theory—here it has been detached from its polemical context.

PSYCHOANALYTIC THEORY:
ID AND EGO, THE UNCONSCIOUS, AND THE INSTINCTS

If I am to say anything intelligible to you, I shall no doubt have to tell you something of a psychological theory which is not known or not appreciated outside analytic circles. It will be easy to deduce from this theory what we want from our patients and how we obtain it. I shall expound it to you dogmatically, as though it were a complete theoretical structure. But do not suppose that it came into being as such a structure, like a philosophical system. We have developed it very slowly, we have wrestled over every small detail of it, we have unceasingly modified it, keeping a continuous contact with observation, till it has finally taken a shape in which it seems to suffice for our purposes. Only a few years ago I should have had to clothe this theory in other terms. Nor, of course, can I guarantee to you that the form in which it is expressed today will remain the final one. Science, as you know, is not a revelation; long after its beginnings it still lacks the attributes of definiteness, immutability and infallibility for which human thought so deeply longs. But such as it is, it is all that we can have. If you will further bear in mind that our science is very young, scarcely as old as the century, and that it is concerned with what is perhaps the most difficult material that can be the subject of human research, you will easily be able to adopt the correct attitude towards my exposition. But interrupt me whenever you feel inclined, if you cannot follow me or if you want further explanations.

"I will interrupt you before you have even begun. You say that you intend to expound a new psychology to me; but I should have thought that psychology was no new science. There have been psychologies and psychologists enough; and I heard of great achievements in that field while I was at college."

I should not dream of disputing them. But if you look into the matter more closely you will have to class these great achievements as belonging rather to the physiology of the sense organs. The theory of mental life could not be developed, because it was inhibited by a single essential misunderstanding. What does it comprise today, as it is taught at college? Apart from those valuable discoveries in the physiology of the senses, a number of classifications and definitions of our mental processes which, thanks to linguistic usage, have become the common property of every educated person. That is clearly not enough to give a view of our mental life. Have you not noticed that every philosopher, every imaginative writer, every historian and every biographer makes up his own psychology for himself, brings forward his own particular hypotheses concerning the interconnections and aims of mental acts—all more or less plausible and all equally untrustworthy? There is an evident lack of any common foundation. And it is for that reason too that in the field of psychology there is, so to speak, no respect and no authority. In that field everyone can "run wild" as he chooses. If you raise a question in physics or chemistry, anyone who knows he possesses no "technical knowledge" will hold his tongue. But if you venture upon a psychological assertion, you must be prepared to meet judgements and contradictions from every quarter. In this field, apparently, there is no "technical knowledge." Everyone has a mental life, so everyone regards himself as a psychologist. But that strikes me as an inadequate legal title. The story is told of how someone who applied for a post as a children's nurse was asked if she knew how to look after babies. "Of course," she replied, "why, after all, I was a baby once myself."

"And you claim that you have discovered this 'common foundation' of mental life, which has been overlooked by every psychologist, from observations on *sick people?*"

The source of our findings does not seem to me to deprive them of their value. Embryology, to take an example, would not deserve to be trusted if it could not give a plain explanation of the origin of innate malformations. I have told you of people whose thoughts go their own way, so that they are obliged to worry over problems to which they are perfectly indifferent. Do you think that academic psychology could ever make the smallest contribution towards explaining an abnormality such as that? And, after all, we all of us have the experience at night-time of our thoughts going their own way and creating things which we do not understand, which puzzle us, and which are suspiciously reminiscent of pathological products. Our dreams, I mean. The common people have always firmly believed that dreams have a sense and a value—that they mean something. Academic psychology has never been able to inform us what this meaning is. It could make nothing of dreams. If it attempted to produce explanations, they were non-psychological—such as tracing them to sensory stimuli, or to an unequal depth of sleep in different portions of the brain, and so on. But it is fair to say that a psychology which cannot explain dreams is also useless for an understanding of normal mental life, that it has no claim to be called a science. . . .

"If I understood rightly, you wanted to tell me about the fundamental postulate of the new psychology."

That was not what I wanted to begin with. My purpose is to let you hear what pictures we have formed of the structure of the mental apparatus in the course of our analytic studies.

"What do you mean by the 'mental apparatus'? and what, may I ask, is it constructed of?"

It will soon be clear what the mental apparatus is; but I must beg you not to ask what material it is constructed of. That is not a subject of psychological interest. Psychology can be as indifferent to it as, for instance, optics can be to the question of whether the walls of a telescope are made of metal or cardboard. We shall leave entirely on one side the *material* line of approach, but not so the *spatial* one. For we picture the unknown apparatus which serves the activities of the mind as being really like an instrument constructed of several parts (which we speak of as "agencies"), each of which performs a particular function and which have a fixed spatial relation to one another: it being understood that by spatial relation—"in front of" and "behind," "superficial" and "deep"—we merely mean in the first instance a representation of the regular succession of the functions. Have I made myself clear?

"Scarcely. Perhaps I shall understand it later. But, in any case, here is a strange anatomy of the soul—a thing which, after all, no longer exists at all for the scientists."

What do you expect? It is a hypothesis like so many others in the sciences: the very earliest ones have always been rather rough. "Open to revision" we can say in such cases. It seems to me unnecessary for me to appeal here to the "as if" which has become so popular. The value of a "fiction" of this kind (as the philosopher Vaihinger would call it) depends on how much one can achieve with its help.

But to proceed. Putting ourselves on the footing of everyday knowledge, we recognize in human beings a mental organization which is interpolated between their sensory stimuli and the perception of their somatic needs on the one hand and their motor acts on the other, and which mediates between them for a particular purpose. We call this organization their *"Ich"* ["ego"; literally, "I"]. Now there is nothing new in this. Each one of us makes this assumption without being a philosopher, and some people even in spite of being philosophers. But this does not, in our opinion, exhaust the description of the mental apparatus. Besides this "I," we recognize another mental region, more extensive, more imposing and more obscure than the "I," and this we call the *"Es"* ["id"; literally, "it"]. The relation between the two must be our immediate concern.

You will probably protest at our having chosen simple pronouns to describe our two agencies or provinces instead of giving them orotund Greek names. In psychoanalysis, however, we like to keep in contact with the popular mode of thinking and prefer to make its concepts scientifically serviceable rather than to reject them. There is no merit in this; we are obliged to take this line; for our theories must be understood by our patients, who are often very intelligent, but not always learned. The impersonal "it" is immediately connected with certain forms of expression used by normal people. "It shot through me," people say; "there was something in me at that moment that was stronger than me." *"C'était plus fort que moi."*

In psychology we can only describe things by the help of analogies. There is nothing peculiar in this; it is the case elsewhere as well. But we have constantly to keep changing these analogies, for none of them lasts us long enough. Accordingly, in trying to make the relation between the ego and the id clear, I must ask you to picture the ego as a kind of façade of the id, as a frontage, like an external, cortical, layer of it. We can hold on to this last analogy. We know that cortical layers owe their peculiar characteristics to the modifying influence of the external medium on which they abut. Thus we suppose that the ego is the layer of the mental apparatus (of the id) which has been modified by the influence of the external world (or reality). This will show you how in psychoanalysis we take spatial ways of looking at things seriously. For us the ego is really something superficial and the id something deeper—looked at from outside, of course. The ego lies between reality and the id, which is what is truly mental.

"I will not ask any questions yet as to how all this can be

known. But tell me first what you gain from this distinction between an ego and an id? What leads you to make it?"

Your question shows me the right way to proceed. For the important and valuable thing is to know that the ego and the id differ greatly from each other in several respects. The rules governing the course of mental acts are different in the ego and the id; the ego pursues different purposes and by other methods. A great deal could be said about this; but perhaps you will be content with a fresh analogy and an example. Think of the difference between "the front" and "behind the lines," as things were during the war. We were not surprised then that some things were different at the front from what they were behind the lines, and that many things were permitted behind the lines which had to be forbidden at the front. The determining influence was, of course, the proximity of the enemy; in the case of the mental life it is the proximity of the external world. There was a time when "outside," "strange" and "hostile" were identical concepts. And now we come to the example. In the id there are no conflicts; contradictions and antitheses persist side by side in it unconcernedly, and are often adjusted by the formation of compromises. In similar circumstances the ego feels a conflict which must be decided; and the decision lies in one urge being abandoned in favour of the other. The ego is an organization characterized by a very remarkable trend towards unification, towards synthesis. This characteristic is lacking in the id; it is, as we might say, "all to pieces"; its different urges pursue their own purposes independently and regardless of one another.

"And if such an important mental region 'behind the lines' exists, how can you explain its having been overlooked till the time of analysis?"

That brings us back to one of your earlier questions. Psychology had barred its own access to the region of the id by insisting on a postulate which is plausible enough but untenable: namely, that all mental acts are conscious to us—that being conscious is the criterion of what is mental and that, if there are processes in our brain which are not conscious, they do not deserve to be called mental acts and are no concern of psychology.

"But I should have thought that was obvious."

Yes, and that is what psychologists think. Nevertheless it can easily be shown to be false—that is, to be a quite inexpedient distinction. The idlest self-observation shows that ideas may occur to us which cannot have come about without preparation. But you experience nothing of these preliminaries of your thought, though they too must certainly have been of a mental nature; all that enters your consciousness is the ready-made result. Occasionally you can make these preparatory thought-structures conscious *in retrospect,* as though in a reconstruction.

"Probably one's attention was distracted, so that one failed to notice the preparations."

Evasions! You cannot in that way get around the fact that acts of a mental nature and often very complicated ones, can take place in you, of which your consciousness learns nothing and of which you know nothing. Or are you prepared to suppose that a greater or smaller amount of your "attention" is enough to transform a non-mental act into a mental one? But what is the use of disputing? There are hypnotic experiments in which the existence of such non-conscious thoughts are irrefutably demonstrated to anyone who cares to learn.

"I shall not retract; but I believe I understand you at last. What you call "ego" is consciousness; and your "id" is the so-called subconscious that people talk about so much nowadays. But why the masquerading with the new names?"

It is not masquerading. The other names are of no use. And do not try to give me literature instead of science. If someone talks of subconsciousness, I cannot tell whether he means the term topographically—to indicate something lying in the mind beneath consciousness—or qualitatively—to indicate another consciousness, a subterranean one, as it were. He is probably not clear about any of it. The only trustworthy antithesis is between conscious and unconscious. But it would be a serious mistake to think that this antithesis coincides with the distinction between ego and id. Of course it would be delightful if it were as simple as that: our theory would have a smooth passage. But things are not so simple. All that is true is that everything that happens in the id is and remains unconscious, and that processes in the ego, and they alone, *can* become conscious. But not all of them are, nor always, nor necessarily; and large portions of the ego can remain permanently unconscious. . . .

"I expect you will want to tell me how, on the basis of the theories of psychoanalysis, the origin of a neurotic illness can be pictured."

I will try to. But for that purpose we must study our ego and our id from a fresh angle, from the *dynamic* one—that is to say, having regard to the forces at work in them and between them. Hitherto we have been content with a *description* of the mental apparatus.

"My only fear is that it may become unintelligible again!"

I hope not. You will soon find your way about in it. Well then, we assume that the forces which drive the mental apparatus into activity are produced in the bodily organs as an expression of the major somatic needs. You will recollect the words of our poet-philosopher: "Hunger and love [are what moves the world]." Incidentally, quite a formidable pair of forces! We give these bodily needs, in so far as they represent an instigation to mental activity, the name of *"Triebe"* [instincts], a word for which we are envied by many modern languages. Well, these instincts fill the id: all the energy in the id, as we may put it briefly, originates from them. Nor have the forces in the ego any other origin; they are derived from those in the id. What, then, do these instincts want? Satisfaction—that is, the establishment of situations in which the bodily needs can be extinguished. A lowering of the tension of need is felt by our organ of consciousness as pleasurable; an increase of it is soon felt as unpleasure. From these oscillations arises the series of feelings of pleasure-unpleasure, in accordance with which the whole mental apparatus regulates its activity. In this connection we speak of a "dominance of the pleasure principle."

If the id's instinctual demands meet with no satisfaction, intolerable conditions arise. Experience soon shows that these situations of satisfaction can only be established with the help of the external world. At that point the portion of the id which is directed towards the external world—the ego—begins to function. If all the driving force that sets the vehicle in motion is derived from the id, the ego, as it were, undertakes the steering, without which no goal can be reached. The instincts in the id press for immediate satisfaction at all costs, and in that way they achieve nothing or even bring about appreciable damage. It is the task of the ego to guard against such mishaps, to mediate between the claims of the id and the objections of the external world. It carries on its activity in two directions. On the one hand, it observes the external world with the help of its sense-organ, the system of consciousness, so as to catch the favourable moment for harmless satisfaction; and on the other hand it influences the id, bridles its "passions," induces its instincts to postpone their satisfaction, and indeed, if the necessity is recognized, to modify its aims, or, in return for some compensation, to give them up. In so far as it tames the id's impulses in this way, it replaces the pleasure principle, which was formerly alone decisive, by what is known as the "reality principle," which, though it pursues the same ultimate aims, takes into account the conditions imposed by the real external world. Later, the ego learns that there is yet another way of securing satisfaction besides the *adaptation* to the external world which I have described. It is also possible to intervene in the external world by *changing* it, and to establish in it intentionally the conditions which make satisfaction possible. This activity then becomes the ego's highest function; decisions as to when it is more expedient to control one's passions and bow before reality, and when it is more expedient to side with them and to take arms against the external world—such decisions make up the whole essence of worldly wisdom.

REPRESSION, NEUROSIS, AND SEXUALITY

"And does the id put up with being dominated like this by the ego, in spite of being, if I understand you aright, the stronger party?"

Yes, all will be well if the ego is in possession of its whole organization and efficiency, if it has access to all parts of the id and can exercise its influence on them. For there is no natural opposition between ego and id; they belong together, and under healthy conditions cannot in practice be distinguished from each other.

"That sounds very pretty; but I cannot see how in such an ideal relation there can be the smallest room for a pathological disturbance."

You are right. So long as the ego and its relations to the id fulfil these ideal conditions, there will be no neurotic disturbance. The point at which the illness makes its breach is an unexpected one, though no one acquainted with general pathology will be surprised to find a confirmation of the principle that it is precisely the most important developments and differentiations that carry in them the seeds of illness, of failure of function.

"You are becoming too learned. I cannot follow you."

I must go back a little bit further. A small living organism is a truly miserable, powerless thing, is it not? compared with the immensely powerful external world, full as it is of destructive influences. A primitive organism, which has not developed any adequate ego-organization, is at the mercy of all these "traumas." It lives by the "blind" satisfaction of its instinctual wishes and often perishes in consequence. The differentiation of an ego is above all a step towards self-preservation. Nothing, it is true, can be learnt from being destroyed; but if one has luckily survived a trauma one takes notice of the approach of similar situations and signalizes the danger by an abbreviated repetition of the impressions one has experienced in connection with the trauma—by an *affect of anxiety*. This reaction to the perception of the danger now introduces an attempt at flight, which can have a life-saving effect till one has grown strong enough to meet the dangers of the external world in a more active fashion—even aggressively, perhaps.

"All this is very far away from what you promised to tell me."

You have no notion how close I am to fulfilling my promise. Even in organisms which later develop an efficient ego-organization, their ego is feeble and little differentiated from their id to begin with, during their first years of childhood. Imagine now what will happen if this powerless ego experiences an instinctual demand from the id which it would already like to resist (because it senses that to satisfy it is dangerous and would conjure up a traumatic situation, a collision with the external world) but which it cannot control, because it does not yet possess enough strength to do so. In such a case the ego treats the instinctual danger as if it was an external one; it makes an attempt at flight, draws back from this portion of the id and leaves it to its fate, after withholding from it all the contributions which it usually makes to instinctual impulses. The ego, as we put it, institutes a *repression* of these instinctual impulses. For the moment this has the effect of fending off the danger; but one cannot confuse the inside and the outside with impunity. One cannot run away from oneself. In repression the ego is following the pleasure principle, which it is usually in the habit of correcting; and it is bound to suffer damage in revenge. This lies in the ego's having permanently narrowed its sphere of influence. The repressed instinctual impulse is now isolated, left to itself, inaccessible, but also uninfluenceable. It goes its own way. Even later, as a rule, when the ego has grown stronger, it still cannot lift the repression; its synthesis is impaired, a part of the id remains forbidden ground to the ego. Nor does the isolated instinctual impulse remain idle; it understands how to make up for being denied normal satisfaction; it produces psychical derivatives which take its place; it links itself to other processes which by its influence it likewise tears away from the ego; and finally it breaks through into the ego and into consciousness in the form of an unrecognizably distorted substitute, and creates what we call a symptom. All at once the nature of a neurotic disorder becomes clear to us: on the one hand an ego which is inhibited in its synthesis, which has no influence on parts of the id, which must renounce some of its activities in order to avoid a fresh collision with what has been repressed, and which exhausts itself in what are for the most part vain acts of defence against the symptoms, the derivatives of the repressed impulses; and on the other hand an id in which individual instincts have made themselves independent, pursue their aims regardless of the interests of the person as a whole and henceforth obey the laws only of the primitive psychology that rules in the depths of the id. If we survey the whole situation we arrive at a simple formula for the origin of a neurosis: the ego has made an attempt to suppress certain portions of the id *in an inappropriate manner,* this attempt has failed and the id has taken its revenge. A neurosis is thus the result of a conflict between the ego and the id, upon which the ego has embarked because, as careful investigation shows, it wishes at all costs to retain its adaptability in relation to the real external world. The disagreement is between the external world and the id; and it is because the ego, loyal to its inmost nature, takes sides with the external world that it becomes involved in a conflict with its id. But please observe that what creates the determinant for the illness is not the fact of this conflict—for disagreements of this kind between reality and the id are unavoidable and it is one of the ego's standing tasks to mediate in them—but the circumstance that the ego has made use of the inefficient instrument of repression for dealing with the conflict. But this in turn is due to the fact that the ego, at the time at which it was set the task, was undeveloped and powerless. The decisive repressions all take place in early childhood.

"What a remarkable business! I shall follow your advice and not make criticisms, since you only want to show me what psycho-analysis believes about the origin of neurosis so that you can go on to say how it sets about combating it. I should have various questions to ask and later on I shall raise some of them. But at the moment I myself feel tempted for once to carry your train of thought further and to venture upon a theory of my own. You have expounded the relation between external world, ego and id, and you have laid it down as the determinant of a neurosis that the ego in its dependence on the external world struggles against the id. Is not the opposite case conceivable of the ego in a conflict of this kind allowing itself to be dragged away by the id and disavowing its regard for the external world? What happens in a case like that? From my lay notions of the nature of insanity I should say that such a decision on the part of the ego might be the determinant of insanity. After all, a turning away of that kind from reality seems to be the essence of insanity."

Yes. I myself have thought of that possibility, and indeed I believe it meets the facts—though to prove the suspicion true would call for a discussion of some highly complicated considerations. Neuroses and psychoses are evidently intimately related, but they must nevertheless differ in some decisive respect. That might well be the side taken by the ego in a conflict of this kind. In both cases the id would retain its characteristic of blind inflexibility.

"Well, go on! What hints on the treatment of neurotic illnesses does your theory give?"

It is easy now to describe our therapeutic aim. We try to restore the ego, to free it from its restrictions, and to give it back the command over the id which it has lost owing to its early repressions. It is for this one purpose that we carry out analysis, our whole technique is directed to this aim. We have to seek out the repressions which have been set up and to urge the ego to correct them with our help and to deal with conflicts better than by an attempt at flight. Since these repressions belong to the very early years of childhood, the work of analysis leads us, too, back to that period. Our path to these situations of conflict, which have for the most part been forgotten and which we try to revive in the patient's memory, is pointed out to us by his symptoms, dreams and free associations. These must, however, first be interpreted—translated—for, under the influence of the psychology of the id, they have assumed forms of expression that are strange to our comprehension. We may assume that whatever associations, thoughts and memories the patient is unable to communicate to us without internal struggles are in some way connected with the repressed material or are its derivatives. By encouraging the patient to disregard his resistances to telling us these things, we are educating his ego to overcome its inclination towards attempts at flight and to tolerate an approach to what is repressed. In the end, if the situation of the repression can be successfully reproduced in his memory, his compliance will be brilliantly rewarded. The whole difference between his age then and now works in his favour; and the thing from which his childish ego fled in terror will often seem to his adult and strengthened ego no more than child's play. . . .

I can now proceed with my exposition. Sexual life is not simply something spicy; it is also a serious scientific problem. There was much that was novel to be learnt about it, many strange things to be explained. I told you just now that analysis has to go back into the early years of the patient's childhood, because the decisive repressions have taken place then, while his ego was feeble. But surely in childhood there is no sexual life? surely it only starts at puberty? On the contrary. We have to learn that sexual instinctual impulses accompany life from birth onwards, and that it is precisely in order to fend off those instincts that the infantile ego institutes repressions. A remarkable coincidence, is it not? that small children should already be struggling against the power of sexuality, just as the speaker in the learned society was to do later, and later still my followers who have set up their own theories. How does that come about? The most general explanation would be that our civilization is built up entirely at the expense of sexuality; but there is much more to be said on the subject.

The discovery of infantile sexuality is one of those of which we have reason to feel ashamed [because of its obviousness]. A few paediatricians have, it seems, always known about it, and a few children's nurses. Clever men, who call themselves child psychologists, have thereupon spoken in tones of reproach of a "desecration of the innocence of childhood." Once again, sentiment instead of argument! Events of that kind are of daily occurrence in political bodies. A member of the Opposition rises and denounces some piece of maladministration in the Civil Service, in the Army, in the Judiciary and so on. Upon this another member, preferably one of the Government, declares that such statements are an affront to the sense of honour of the body politic, of the army, of the dynasty, or even of the nation. So they are as good as untrue. Feelings such as these can tolerate no affronts.

The sexual life of children is of course different from that of adults. The sexual function, from its beginnings to the definitive form in which it is so familiar to us, undergoes a complicated process of development. It grows together from numerous component instincts with different aims and passes through several phases of organization till at last it comes into the service of reproduction. Not all the component instincts are equally serviceable for the final outcome; they must be diverted, remodelled and in part suppressed. Such a far-reaching course of development is not always passed through without a flaw; inhibitions in development take place, partial fixations at early stages of development. If obstacles arise later on to the exercise of the sexual function, the sexual urge—the libido, as we call it—is apt to hark back to these earlier points of fixation. The study of the sexuality of children and its transformations up to maturity has also given us the key to understanding of what are known as the sexual perversions, which people used always to describe with all the requisite indications of disgust but whose origin they were never able to explain. The whole topic is of uncommon interest, but for the purposes of our conversation there is not much sense in telling you more about it. To find one's way about in it one of course needs anatomical and physiological knowledge, all of which is unfortunately not to be acquired in medical schools. But a familiarity with the history of civilization and with mythology is equally indispensable.

"After all that, I still cannot form any picture of the sexual life of children."

Then I will pursue the subject further; in any case it is not easy for me to get away from it. I will tell you, then, that the most remarkable thing about the sexual life of children seems to me that it passes through the whole of its very far-reaching development in the first five years of life. From then onwards until puberty there stretches what is known as the period of latency. During it sexuality normally advances no further; on the contrary, the sexual urges diminish in strength and many things are given up and forgotten which the child did and knew. During that period of life, after the efflorescence of sexuality has withered, such attitudes of the ego as shame, disgust and morality arise, which are destined to stand up against the later tempest of puberty and to lay down the path of the freshly awakening sexual desires. This "diphasic onset," as it is named, of sexual life has a great deal to do with the genesis of neurotic illnesses. It seems to occur only in human beings, and it is perhaps one of the determinants of the human privilege of becoming neurotic. The prehistory of sexual life was just as much overlooked before psychoanalysis as, in another department, the background to conscious mental life. You will rightly suspect that the two are intimately connected.

There is much to be told, for which our expectations have not prepared us, about the contents, manifestations and achievements of this early period of sexuality. For instance, you will no doubt be surprised to hear how often little boys are afraid of being eaten up by their father. (And you may also be surprised at my including this fear among the phenomena of sexual life.) But I may remind you of the mythological tale which you may still recall from your schooldays of how the god Kronos swallowed his children. How strange this must have sounded to you when you first heard it! But I suppose none of us thought about it at the time. To-day we can also call to mind a number of fairy tales in which some ravenous animal like a wolf appears, and we shall recognize it as a disguise of the father. And this is an opportunity of assuring you that it was only through the knowledge of infantile sexuality that it became possible to understand mythology and the world of fairy tales. Here then something has been gained as a by-product of analytic studies.

You will be no less surprised to hear that male children suffer from a fear of being robbed of their sexual organ by their father, so that this fear of being castrated has a most powerful influence on the development of their character and in deciding the direction to be followed by their sexuality. And here again mythology may give you the courage to believe psychoanalysis. The same Kronos who swallowed his children also emasculated his father Uranus, and was afterwards himself emasculated in revenge by his son Zeus, who had been rescued through his mother's cunning. If you have felt inclined to suppose that all that psychoanalysis reports about the early sexuality of children is derived from the disordered imagination of the analysts, you must at least admit that their imagination has created the same product as the imaginative activities of primitive man, of which myths and fairy tales are the precipitate. The alternative friendlier, and probably also the more pertinent view would be that in the mental life of children to-day we can still detect the same archaic factors which were once dominant generally in the primaeval days of human civilization. In his mental development the child would be repeating the history of his race in an abbreviated form, just as embryology long since recognized was the case with somatic development.

Another characteristic of early infantile sexuality is that the female organ proper as yet plays no part in it: the child has not yet discovered it. Stress falls entirely on the male organ, all the child's interest is directed towards the questions of whether it is present or not. We know less about the sexual life of little girls than of boys. But we need not feel ashamed of this distinction; after all, the sexual life of adult women is a "dark continent" for psychology. But we have learnt that girls feel deeply their lack of a sexual organ that is equal in value to the male one; they regard themselves on that account as inferior, and this "envy for the penis" is the origin of a whole number of characteristic feminine reactions.

It is also characteristic of children that their two excretory needs are cathected [charged] with sexual interest. Later on, education draws a sharp distinction here, which is once more obliterated in the practice of joking. It may seem to us an unsavoury fact, but it takes quite a long time for children to develop feelings of disgust. This is not disputed even by people who insist otherwise on the seraphic purity of the child's mind.

Nothing, however, deserves more notice than the fact that chil-

dren regularly direct their sexual wishes towards their nearest relatives—in the first place, therefore, towards their father and mother,
and afterwards towards their brothers and sisters. The first object of
a boy's love is his mother, and of a girl's her father (except in so far
as an innate bisexual disposition favours the simultaneous presence
of the contrary attitude). The other parent is felt as a disturbing
rival and not infrequently viewed with strong hostility. You must
understand me aright. What I mean to say is not that the child
wants to be treated by its favourite parent merely with the kind of
affection which we adults like to regard as the essence of the parent-
child relation. No, analysis leaves us in no doubt that the child's
wishes extend beyond such affection to all that we understand by
sensual satisfaction—so far, that is, as the child's powers of imagination allow. It is easy to see that the child never guesses the actual
facts of sexual intercourse; he replaces them by other notions derived from his own experience and feelings. As a rule his wishes
culminate in the intention to bear, or in some indefinable way, to
procreate a baby. Boys, too, in their ignorance, do not exclude
themselves from the wish to bear a baby. We give the whole of this
mental structure the name of "Oedipus complex," after the familiar
Greek legend. With the end of the early sexual period it should
normally be given up, should radically disintegrate and become
transformed; and the results of this transformation are destined for
important functions in later mental life. But as a rule this is not
effected radically enough, in which case puberty brings about a
revival of the complex, which may have serious consequences.

I am surprised that you are still silent. That can scarcely mean
consent.—In asserting that a child's first choice of an object is, to use
the technical term, an incestuous one, analysis no doubt once more
hurt the most sacred feelings of humanity, and might well be prepared for a corresponding amount of disbelief, contradiction and
attack. And these it has received in abundance. Nothing has damaged
it more in the good opinion of its contemporaries than its hypothesis
of the Oedipus complex as a structure universally bound to human
destiny. The Greek myth, incidentally, must have had the same
meaning; but the majority of men to-day, learned and unlearned
alike, prefer to believe that Nature has laid down an innate abhorrence in us as a guard against the possibility of incest. . . .

PSYCHOANALYTIC TREATMENT:
RESISTANCE, THE SUPEREGO, AND TRANSFERENCE

"I believe I understand your purpose. You want to show me what kind of knowledge is needed in order to practise analysis, so that I may be able to judge whether only doctors should have a right to do so. Well, so far very little to do with medicine has turned up: a great deal of psychology and a little biology or sexual science. But perhaps we have not got to the end?"

Decidedly not. There are still gaps to be filled. May I make a request? Will you describe how you now picture an analytic treatment?—just as though you had to undertake one yourself.

"A fine idea, to be sure! No, I have not the least intention of settling our controversy by an experiment of that sort. But just to oblige, I will do what you ask—the responsibility will be yours. Very well. I will suppose that the patient comes to me and complains of his troubles. I promise him recovery or improvement if he will follow my directions. I call on him to tell me with perfect candour everything that he knows and that occurs to him, and not to be deterred from that intention even if some things are disagreeable to say. Have I taken in the rule properly?"

Yes. You should add: "even if what occurs to him seems unimportant or senseless."

"I will add that. Thereupon he begins to talk and I listen. And what then? I infer from what he tells me the kind of impressions, experiences and wishes which he has repressed because he came across them at a time when his ego was still feeble and was afraid of them instead of dealing with them. When he has learnt this from me, he puts himself back in the old situations and with my help he manages better. The limitations to which his ego was tied then disappear, and he is cured. Is that right?"

Bravo! bravo! I see that once again people will be able to accuse me of having made an analyst of someone who is not a doctor. You have mastered it all admirably.

"I have done no more than repeat what I have heard from you—as though it was something I had learnt by heart. All the same, I cannot form any picture of how I should do it, and I am quite at a loss to understand why a job like that should take an hour a day for so many months. After all, an ordinary person has not as a rule experienced such a lot, and what was repressed in childhood is probably in every case the same."

When one really practises analysis one learns all kinds of things besides. For instance: you would not find it at all such a simple matter to deduce from what the patient tells you the experiences he has forgotten and the instinctual impulses he has repressed. He says something to you which at first means as little to you as it does to him. You will have to make up your mind to look at the material which he delivers to you in obedience to the rule in a quite special way: as though it were ore, perhaps, from which its content of precious metal has to be extracted by a particular process. You will be prepared, too, to work over many tons of ore which may contain but little of the valuable material you are in search of. Here we should have a first reason for the prolonged character of the treatment.

"But how does one work over this raw material—to keep to your simile?"

By assuming that the patient's remarks and associations are only distortions of what you are looking for —allusions, as it were, from which you have to guess what is hidden behind them. In a word, this material, whether it consists of memories, associations or dreams, has first to be *interpreted*. You will do this, of course, with an eye to the expectations you have formed as you listened, thanks to your special knowledge.

" 'Interpret!' A nasty word! I dislike the sound of it; it robs me of all certainty. If everything depends on my interpretation who can guarantee that I interpret right? So after all everything *is* left to my caprice."

Just a moment! Things are not quite as bad as that. Why do you choose to except your own mental processes from the rule of law which you recognize in other people's? When you have attained some degree of self-discipline and have certain knowledge at your disposal, your interpretations will be independent of your personal characteristics and will hit the mark. I am not saying that the analyst's personality is a matter of indifference for this portion of his task. A kind of sharpness of hearing for what is unconscious and repressed, which is not possessed equally by everyone, has a part to play. And here, above all, we are brought to the analyst's obligation to make himself capable, by a deep-going analysis of his own, of the unprejudiced reception of the analytic material. Something, it is true, still remains over: something comparable to the "personal equation" in astronomical observations. This individual factor will always play a larger part in psychoanalysis than elsewhere. An abnormal person can become an accurate physicist; as an analyst he will be hampered by his own abnormality from seeing the pictures of mental life undistorted. Since it is impossible to demonstrate to anyone his own abnormality, general agreement in matters of depth-psychology will be particularly hard to reach. Some psychologists, indeed, think it is quite impossible and that every fool has an equal right to give out his folly as wisdom. I confess that I am more of an optimist about this. After all, our experiences show that fairly satisfactory agreements can be reached even in psychology. Every field of research has its particular difficulty which we must try to eliminate. And, moreover, even in the interpretative art of analysis there is much that can be learnt like any other material of study: for instance, in connection with the peculiar method of indirect representation through symbols.

"Well, I no longer have any desire to undertake an analytic treatment even in my imagination. Who can say what other surprises I might meet with?"

You are quite right to give up the notion. You see how much more training and practice would be needed. When you have found the right interpretation, another task lies ahead. You must wait for the right moment at which you can communicate your interpretation to the patient with some prospect of success.

"How can one always tell the right moment?"

That is a question of tact, which can become more refined with experience. You will be making a bad mistake if, in an effort, perhaps, at shortening the analysis, you throw your interpretations at the patient's head as soon as you have found them. In that way you will draw expressions of resistance, rejection and indignation from him; but you will not enable his ego to master his repressed material. The formula is: to wait till he has come so near to the repressed material that he has only a few more steps to take under the lead of the interpretation you propose.

"I believe I should never learn to do that. And if I carry out these precautions in making my interpretation, what next?"

It will then be your fate to make a discovery for which you were not prepared.

"And what may that be?"

That you have been deceived in your patient; that you cannot count in the slightest on his collaboration and compliance; that he is ready to place every possible difficulty in the way of your common work—in a word, that he has no wish whatever to be cured.

"Well! that is the craziest thing you have told me yet. And I do not believe it either. The patient who is suffering so much, who complains so movingly about his troubles, who is making so great a sacrifice for the treatment—you say he has no wish to be cured! But of course you do not mean what you say."

Calm yourself! I *do* mean it. What I said was the truth—not the whole truth, no doubt, but a very noteworthy part of it. The patient wants to be cured—but he also wants not to be. His ego has lost its unity, and for that reason his will has no unity either. If that were not so, he would be no neurotic.

" 'Were I sagacious, I should not be Tell!' "

The derivatives of what is repressed have broken into his ego and established themselves there; and the ego has as little control over trends from that source as it has over what is actually repressed, and as a rule it knows nothing about them. These patients, indeed, are of a peculiar nature and raise difficulties with which we are not accustomed to reckon. All our social institutions are framed for people with a united and normal ego, which one can classify as good or bad, which either fulfils its function or is altogether eliminated by an overpowering influence. Hence the juridical alternative: responsible or irresponsible. None of these distinctions apply to neurotics. It must be admitted that there is difficulty in adapting social demands to their psychological condition. This was experienced on a large scale during the last war. Were the neurotics who evaded service malingerers or not? They were both. If they were treated as malingerers and if their illness was made highly uncomfortable, they recovered; if after being ostensibly restored they were sent back into service, they promptly took flight once more into illness. Nothing could be done with them. And the same is true of neurotics in civil life. They complain of their illness but exploit it with all their strength; and if someone tries to take it away from them they defend it like the proverbial lioness with her young. Yet there would be no sense in reproaching them for this contradiction.

"But would not the best plan be not to give these difficult people any treatment at all, but to leave them to themselves? I cannot think it is worth while to expend such great efforts over each of them as you lead me to suppose that you make."

I cannot approve of your suggestion. It is undoubtedly a more proper line to accept the complications of life rather than struggle against them. It may be true that not every neurotic whom we treat is worth the expenditure of an analysis; but there are some very valuable individuals among them as well. We must set ourselves the goal of bringing it about that as few human beings as possible enter civilized life with such a defective mental equipment. And for that purpose we must collect much experience and learn to understand many things. Every analysis can be instructive and bring us a yield of new understanding quite apart from the personal value of the individual patient.

"But if a volitional impulse has been formed in the patient's ego which wishes to retain the illness, it too must have its reasons and motives and be able in some way to justify itself. But it is impossible to see why anyone should want to be ill or what he can get out of it."

Oh, that is not so hard to understand. Think of the war neurotics, who do not have to serve, precisely because they are ill. In civil life illness can be used as a screen to gloss over incompetence in one's profession or in competition with other people; while in the family it can serve as a means for sacrificing the other members and extorting proofs of their love or for imposing one's will upon them. All of this lies fairly near the surface; we sum it up in the term "gain from illness." It is curious, however, that the patient—that is, his ego—nevertheless knows nothing of the whole concatenation of these motives and the actions which they involve. One combats the influence of these trends by compelling the ego to take cognizance of them. But there are other motives, that lie still deeper, for holding on to being ill, which are not so easily dealt with. But these cannot be understood without a fresh journey into psychological theory.

"Please go on. A little more theory will make no odds now."

When I described the relation between the ego and the id to you, I suppressed an important part of the theory of the mental apparatus. For we have been obliged to assume that within the ego itself a particular agency has become differentiated, which we name the super-ego. This super-ego occupies a special position between the ego and the id. It belongs to the ego and shares its high degree of psychological organization; but it has a particularly intimate connection with the id. It is in fact a precipitate of the first object-cathexes of the id and is the heir to the Oedipus complex after its demise. This super-ego can confront the ego and treat it like an object; and it often treats it very harshly. It is as important for the ego to remain on good terms with the super-ego as with the id. Estrangements between the ego and the super-ego are of great significance in mental life. You will already have guessed that the super-ego is the vehicle of the phenomenon that we call conscience. Mental health very much depends on the super-ego's being normally developed—that is, on its having become sufficiently impersonal. And that is precisely what it is not in neurotics, whose Oedipus complex has not passed through the correct process of transformation. Their super-ego still confronts their ego as a strict father confronts a child; and their morality operates in a primitive fashion in that the ego gets itself punished by the super-ego. Illness is employed as an instrument for this "self-punishment," and neurotics have to behave as though they were governed by a sense of guilt which, in order to be satisfied, needs to be punished by illness.

"That really sounds most mysterious. The strangest thing about it is that apparently even this mighty force of the patient's conscience does not reach his consciousness."

Yes, we are only beginning to appreciate the significance of all these important circumstances. That is why my description was bound to turn out so obscure. But now I can proceed. We describe all the forces that oppose the work of recovery as the patient's "resistances." The gain from illness is one such resistance. The "unconscious sense of guilt" represents the super-ego's resistance; it is the most powerful factor, and the one most dreaded by us. We meet with still other resistances during the treatment. If the ego during the early period has set up a repression out of fear, then the fear still persists and manifests itself as a resistance if the ego approaches the repressed material. And finally, as you can imagine, there are likely to be difficulties if an instinctual process which has been going along a particular path for whole decades is suddenly expected to take a new path that has just been made open for it. That might be called the id's resistance. The struggle against all these resistances is our main work during an analytic treatment; the task of making interpretations is nothing compared to it. But as a result of this struggle and of the overcoming of the resistances, the patient's ego is so much altered and strengthened that we can look forward calmly to his future behaviour when the treatment is over. On the other hand, you can understand now why we need such long treatments. The length of the path of development and the wealth of material are not the decisive factors. It is more a question of whether the path is clear. Any army can be held up for weeks on a stretch of country which in peace time an express train crosses in a couple of hours—if the army has to overcome the enemy's resistance there. Such battles call for time in mental life too. I am unfortunately obliged to tell you that every effort to hasten analytic treatment appreciably has hitherto failed. The best way of shortening if seems to be to carry it out according to the rules.

"If I ever felt any desire to poach on your preserves and try my hand at analysing someone else, what you tell me about the resistances would have cured me of it. But how about the special personal influence that you yourself have after all admitted? Does not that come into action against the resistances?"

It is a good thing you have asked me about that. This personal influence is our most powerful dynamic weapon. It is the new element which we introduce into the situation and by means of which we make it fluid. The intellectual content of our explanations cannot do it, for the patient, who shares all the prejudices of the world around him, need believe us as little as our scientific critics do. The neurotic sets to work because he has faith in the analyst, and he believes him because he acquires a special emotional attitude towards the figure of the analyst. Children, too, only believe people they are attached to. I have already told you what use we make of this particularly large "suggestive" influence. Not for suppressing the symptoms—this distinguishes the analytic method from other psychotherapeutic procedures—but as a motive force to induce the patient to overcome his resistances.

"Well, and if that succeeds, does not everything then go smoothly?"

Yes, it ought to. But there turns out to be an unexpected complication. It was perhaps the greatest of the analyst's surprises to find that the emotional relation which the patient adopts towards him is of a quite peculiar nature. The very first doctor who attempted an analysis—it was not myself—came up against this phenomenon and did not know what to make of it. For this emotional relation is, to put it plainly, in the nature of falling in love. Strange, is it not? Especially when you take into account that the analyst does nothing to provoke it but on the contrary rather keeps at a distance from the patient, speaking humanly, and surrounds himself with some degree of reserve—when you learn besides that this odd love-relationship disregards anything else that is really propitious and every variation in personal attraction, age, sex or class. This love is of a positively compulsive kind. Not that that characteristic need be absent from spontaneous falling in love. As you know, the contrary is often the case. But in the analytic situation it makes its appearance with complete regularity without there being any rational explanation for it. One would have thought that the patient's relation to the analyst called for no more than a certain amount of respect, trust, gratitude and human sympathy. Instead, there is this falling in love, which itself gives the impression of being a pathological phenomenon.

"I should have thought all the same that it would be favourable for your analytic purposes. If someone is in love, he is amenable, and he will do anything in the world for the sake of the other person."

Yes. It *is* favourable to start with. But when this falling in love has grown deeper, its whole nature comes to light, much of which is incompatible with the task of analysis. The patient's love is not satisfied with being obedient; it grows exacting, calls for affection-ate and sensual satisfactions, it demands exclusiveness, it develops jealousy, and it shows more and more clearly its reverse side, its readiness to become hostile and revengeful it if cannot obtain its ends. At the same time, like all falling in love, it drives away all other mental material; it extinguishes interest in the treatment and in recovery—in short, there can be no doubt that it has taken the place of the neurosis and that our work has had the result of driving out one form of illness with another.

"That does sound hopeless! What can be done about it? The analysis would have to be given up. But if, as you say, the same thing happens in every case, it would be impossible to carry through any analyses at all."

We will begin by using the situation in order to learn something from it. What we learn may then perhaps help us to master it. Is it not an extremely noteworthy fact that we succeed in transforming every neurosis, whatever its content, into a condition of pathologi-cal love?

Our conviction that a portion of erotic life that has been abnor-mally employed lies at the basis of neuroses must be unshakeably strengthened by this experience. With this discovery we are once more on a firm footing and can venture to make this love itself the object of analysis. And we can make another observation. Analytic love is not manifested in every case as clearly and blatantly as I have tried to depict it. Why not? We can soon see. In proportion as the purely sensual and the hostile sides of his love try to show them-selves, the patient's opposition to them is aroused. He struggles against them and tries to repress them before our very eyes. And now we understand what is happening. The patient is *repeating* in the form of *falling in love* with the analyst mental experiences which he has already been through once before; he has *transferred* on to the analyst mental attitudes that were lying ready in him and were intimately connected with his neurosis. He is also repeating before our eyes his old defensive actions; he would like best to repeat in his relation to the analyst *all* the history of that forgotten period in his life. So what he is showing us is the kernel of his intimate life history: *he is reproducing it tangibly, as though it were actually happening, instead of remembering it.* In this way the riddle of the transference-love is solved and the analysis can proceed on its way—with the *help* of the new situation which had seemed such a menace to it.

"That is very cunning. And is the patient so easy to convince that he is not in love but only obliged to stage a revival of an old piece?"

Everything now depends on that. And the whole skill in handling the "transference" is devoted to bringing it about. As you see, the requirements of analytic technique reach their maximum at this point. Here the gravest mistakes can be made or the greatest successes be registered. It would be folly to attempt to evade the difficulties by suppressing or neglecting the transference; whatever else had been done in the treatment, it would not deserve the name of analysis. To send the patient away as soon as the inconveniences of his transference-neurosis make their appearance would be no more sensible, and would moreover be cowardly. It would be as though one had conjured up spirits and run away from them as soon as they appeared. Sometimes, it is true, nothing else is possible. There are cases in which one cannot master the unleashed transference and the analysis has to be broken off; but one must at least have struggled with the evil spirits to the best of one's strength. To yield to the demands of the transference, to fulfil the patient's wishes for affectionate and sensual satisfaction, is not only justly forbidden by moral considerations but is also completely ineffective as a technical method for attaining the purpose of the analysis. A neurotic cannot be cured by being enabled to reproduce uncorrected an unconscious stereotype plate that is ready to hand in him. If one engages in compromises with him by offering him partial satisfactions in exchange for his further collaboration in the analysis, one must beware of falling into the ridiculous situation of the cleric who was supposed to convert a sick insurance agent. The sick man remained unconverted but the cleric took his leave insured. The only possible way out of the transference situation is to trace it back to the patient's past, as he really experienced it or as he pictured it through the wish-fulfilling activity of his imagination. And this demands from the analyst much skill, patience, calm and self-abnegation. . . .

THE SOCIAL APPLICATIONS OF PSYCHOANALYSIS

We do not consider it at all desirable for psychoanalysis to be swallowed up by medicine and to find its last resting-place in a text-book of psychiatry under the heading "Methods of Treatment," alongside of procedures such as hypnotic suggestion, autosuggestion, and persuasion, which, born from our ignorance, have to thank the laziness and cowardice of mankind for their short-lived effects. It deserves a better fate and, it may be hoped, will meet with one. As a "depth-psychology," a theory of the mental unconscious, it can become indispensable to all the sciences which are concerned with the evolution of human civilization and its major institutions such as art, religion and the social order. It has already, in my opinion, afforded these sciences considerable help in solving their problems. But these are only small contributions compared with what might be achieved if historians of civilization, psychologists of religion, philologists and so on would agree themselves to handle the new instrument of research which is at their service. The use of analysis for the treatment of the neuroses is only one of its applications; the future will perhaps show that it is not the most important one. In any case it would be wrong to sacrifice all the other applications to this single one, just because it touches on the circle of medical interests.

For here a further prospect stretches ahead, which cannot be encroached upon with impunity. If the representatives of the various mental sciences are to study psychoanalysis so as to be able to apply its methods and angles of approach to their own material, it will not be enough for them to stop short at the findings which are laid down in analytic literature. They must learn to understand analysis in the only way that is possible—by themselves undergoing an analysis. The neurotics who need analysis would thus be joined by a second class of persons, who accept analysis from intellectual motives, but who will no doubt also welcome the increase in their capacities which they will incidentally achieve. To carry out these analyses a number of analysts will be needed, for whom any medical knowledge will have particularly little importance. But these "teaching analysts"—let us call them—will require to have had a particularly careful education. If this is not to be stinted, they must be given an opportunity of collecting experience from instructive and informative cases; and since healthy people who also lack the motive of curiosity do not present themselves for analysis, it is once more only upon neurotics that it will be possible for the teaching analysts— under careful supervision—to be educated for their subsequent non-medical activity. All this, however, requires a certain amount of freedom of movement, and is not compatible with petty restrictions.

Perhaps you do not believe in these purely theoretical interests of psychoanalysis or cannot allow them to affect the practical question of lay analysis. Then let me advise you that psychoanalysis has yet another sphere of application, which is outside the scope of the quackery law and to which the doctors will scarcely lay claim. Its application, I mean, to the bringing-up of children. If a child begins to show signs of an undesirable development, if it grows moody, refractory and inattentive, the paediatrician and even the school doctor can do nothing for it, even if the child produces clear neurotic symptoms, such as nervousness, loss of appetite, vomiting or insomnia. A treatment that combines analytic influence with educational measures, carried out by people who are not ashamed to concern themselves with the affairs in a child's world, and who understand how to find their way into a child's mental life, can bring about two things at once: the removal of the neurotic symptoms and the reversal of the change in character which had begun. Our recognition of the importance of these inconspicuous neuroses of children as laying down the disposition for serious illnesses in later life points to these child analyses as an excellent method of prophylaxis. Analysis undeniably still has its enemies. I do not know whether they have means at their command for stopping the activities of these educational analysts or analytic educationalists. I do not think it very likely; but one can never feel too secure.

Moreover, to return to our question of the analytic treatment of adult neurotics, even there we have not yet exhausted every line of approach. Our civilization imposes an almost intolerable pressure on us and it calls for a corrective. Is it too fantastic to expect that psychoanalysis in spite of its difficulties may be destined to the task of preparing mankind for such a corrective?

J. B. Watson

The American psychologist John B. Watson has earned a lasting place in the history of psychology by founding and vigorously championing the behaviorist movement. When Hume talked of applying the methods of science to human nature, he assumed a sharp distinction between the material world of bodies and the mental world of ideas. And when Mill talked of the science of psychology, what he meant, as we have already seen, was the study of laws that would relate introspectable events such as thoughts, emotions, and sensations. Toward the end of the nineteenth century, people such as Wundt and the others Watson mentions tried to establish psychology as an experimental science, seeking to find just such laws as Mill envisaged. Freud, Watson's contemporary, extended the range of the term "mental" to cover states and processes that are not available to the consciousness of their owner, but he too assumed that introspectable phenomena such as dreams were *among* the data that any psychological theory had to explain.

Thus Watson was able to make a splash by redefining the very subject matter of psychology as the outward, publicly observable behavior of living creatures. In his enthusiasm he rejected not only Platonic-Cartesian dualism (which is not so essentially connected with religion as he thought), but also what seems to be the ordinary common sense notion of consciousness as a stream of thoughts, desires, sensations, and emotions. He boldly claims here to be able to show contradictions in these familiar mental concepts.

Watson's crusading vision of a new psychology, couched purely in terms of stimulus and response (where the latter term is to cover every bodily movement, from intentional activity such as writing books to more automatic reflexes), is taken up by B. F. Skinner, as is his rather disturbing interest in not merely explaining, but controlling, our behavior. This excerpt is from Chapter 1 of *Behaviorism* (W. W. Norton, 1st ed., 1925).

THE BEHAVIORIST DEFINITION OF PSYCHOLOGY

Before beginning our study of "behaviorism" or "behavioristic" psychology, it will be worth our while to take a few minutes to

look at the conventional school of psychology that flourished before the advent of behaviorism in 1912—and that still flourishes.
Indeed we should point out at once that behaviorism has not as yet
by any means replaced the older psychology—called *introspective
psychology*—of James, Wundt, Külpe, Titchener, Angell, Judd, and
McDougall. Possibly the easiest way to bring out the contrast between the old psychology and the new is to say that all schools of
psychology except that of behaviorism claim that *"consciousness" is
the subject matter of psychology*. Behaviorism, on the contrary, holds
that the subject matter of human psychology is the *behavior or activities of the human being*. Behaviorism claims that "consciousness" is
neither a definable nor a usable concept; that it is merely another
word for the "soul" of more ancient times. The old psychology is
thus dominated by a kind of subtle religious philosophy. . . .

One example of such a concept is that there is a fearsome God and
that every individual has a soul which is separate and distinct from
the body. This soul is really a part of the supreme being. This concept has led to the philosophical platform called "dualism." All psychology except behaviorism is dualistic. That is to say we have both
a mind (soul) and a body. This dogma has been present in human
psychology from earliest antiquity. No one has ever touched a soul,
or has seen one in a test tube, or has in any way come into relationship with it as he has with the other objects of his daily experience.
Nevertheless, to doubt its existence is to become a heretic and once
might possibly even have led to the loss of one's head. Even today
the man holding a public position dare not question it.

With the development of the physical sciences which came with
the renaissance, a certain release from this stifling soul cloud was
obtained. A man could think of astronomy, of the celestial bodies
and their motions, of gravitation and the like, without involving
soul. Although the early scientists were as a rule devout Christians,
nevertheless they began to leave soul out of their test tubes. Psychology and philosophy, however, in dealing as they thought with non-
material objects, found it difficult to escape the language of the
church, and hence the concepts of mind and soul come down to the
latter part of the nineteenth century. It was the boast of Wundt's
students, in 1879, when the first psychological laboratory was established, that psychology had at last become a science without a soul.

For fifty years we have kept this pseudo-science, exactly as Wundt laid it down. All that Wundt and his students really accomplished was to substitute for the word "soul" the word "consciousness."

From the time of Wundt on, consciousness becomes the keynote of psychology. It is the keynote of all psychologies today except behaviorism. It is a plain assumption just as unprovable, just as unapproachable as the old concept of the soul. And to the behaviorist the two terms are essentially identical, so far as concerns their metaphysical implications.

To show how unscientific is the concept, look for a moment at William James' definition of psychology. "Psychology is the description and explanation of states of consciousness as such." Starting with a definition which *assumes* what he starts out to prove he escapes his difficulty by an *argumentum ad hominem*. Consciousness—Oh, yes, everybody must know what this "consciousness" is. When we have a sensation of red, a perception, a thought, when we *will* to do something, or when we *purpose* to do something, or when we desire to do something, we are being *conscious*. All other introspectionists are equally illogical. In other words, they do not tell us what consciousness is, but merely begin to put things into it by assumption; and then when they come to analyze consciousness, naturally they find in it just what they put into it. Consequently, in the analyses of consciousness made by certain of the psychologists you find such elements as *sensations* and their ghosts, the *images*. With others you find not only sensations, but so-called *affective elements;* in still others you find such elements as *will* —the so-called conative element in consciousness. With some psychologists you find many hundreds of sensations of a certain type; others maintain that only a few of that type exist. And so it goes. Literally hundreds of thousands of printed pages have been published on the minute analysis of this intangible something called "consciousness." And how do we begin work upon it? Not by analyzing it as we would a chemical compound, or the way a plant grows. No, those things are material things. This thing we call consciousness can be analyzed only by *introspection*—a looking in on what goes on inside of us.

As a result of this major assumption that there is such a thing as consciousness and that we can analyze it by introspection, we find as many analyses as there are individual psychologists. There is no way of experimentally attacking and solving psychological problems and standardizing methods.

In 1912 the behaviorists reached the conclusion that they could no longer be content to work with intangibles and unapproachables. They decided either to give up psychology or else to make it a natural science. They saw their brother-scientists making progress in medicine, in chemistry, in physics. Every new discovery in those fields was of prime importance; every new element isolated in one laboratory could be isolated in some other laboratory; each new element was immediately taken up in the warp and woof of science as a whole. May I call your attention to the wireless, to radium, to insulin, to thyroxin, and hundreds of others? Elements so isolated and methods so formulated immediately began to function in human achievement.

In his first efforts to get uniformity in subject matter and in methods the behaviorist began his own formulation of the problem of psychology by sweeping aside all mediaeval conceptions. He dropped from his scientific vocabulary all subjective terms such as sensation, perception, image, desire, purpose, and even thinking and emotion as they were subjectively defined.

The behaviorist asks: Why don't we make what we can *observe* the real field of psychology? Let us limit ourselves to things that can be observed, and formulate laws concerning only those things. Now what can we observe? Well, we can observe *behavior—what the organism does or says.* And let me make this fundamental point at once: that *saying* is doing—that is, *behaving.* Speaking overtly or to ourselves (thinking) is just as objective a type of behavior as baseball.

The rule, or measuring rod, which the behaviorist puts in front of him always is: Can I describe this bit of behavior I see in terms of "stimulus and response."? By stimulus we mean any object in the general environment or any change in the tissues themselves due to the physiological condition of the animal, such as the change we get when we keep an animal from sex activity, when we keep it from feeding, when we keep it from building a nest. By response we mean anything the animal does—such as turning towards or away from a light, jumping at a sound, and more highly organized activities such as building a skyscraper, drawing plans, having babies, writing books, and the like. . . .

After hearing this brief survey of the behavioristic approach to the problems of psychology, I can almost hear you exclaiming: "Why, yes, it is worth while to study human behavior in this way, but the study of behavior is not the whole of psychology. It leaves out too much. Don't I have sensations, perceptions, conceptions? Do I not forget things and remember things, imagine things, have visual images and auditory images of things I once have seen and heard? Can I not see and hear things that I have never seen or heard in nature? Can I not be attentive or inattentive? Can I not will to do a thing or will not to do it, as the case may be? Do not certain things arouse pleasure in me, and others displeasure? Behaviorism is trying to rob us of everything we have believed in since earliest childhood."

Having been brought up on introspective psychology, as most of you have, these questions are perfectly natural and you will find it hard to put away this terminology and begin to formulate your psychological life in terms of behaviorism. Behaviorism is new wine and it will not go into old bottles; therefore I am going to try to make new bottles out of you. I am going to ask you to put away all of your old presuppositions and to allay your natural antagonism and accept the behavioristic platform at least for this series of lectures. Before they end I hope you will find that you have progressed so far with behaviorism that the questions you now raise will answer themselves in a perfectly satisfactory natural science way. Let me hasten to add that if I were to ask you to tell me what you mean by the terms you have been in the habit of using I could soon make you tongue tied with contradictions. I believe I could even convince you that you do not know what you mean by them. You have been using them uncritically as a part of your social and literary tradition. Let us forget them until later lectures.

This is the fundamental starting point of behaviorism. You will soon find that instead of self-observation being the easiest and most natural way of studying psychology, it is an impossible one; you can observe in yourselves only the most elementary forms of response. You will find, on the other hand, that when you begin to study what your neighbor is doing, you will rapidly become proficient in giving a reason for his behavior and in setting situations, (presenting stimuli) that will make him behave in a predictable manner.

Definitions are not as popular today as they used to be. The definition of any one science, physics, for example, would necessarily include the definition of all other sciences. And the same is true of behaviorism. About all that we can do in the way of defining a science at the present time is to mark a ring around that part of the whole of natural science that we claim particularly as our own.

Behaviorism, as you have already grasped from our preliminary discussion, is, then, a natural science that takes the whole field of human adjustments as its own. Its closest scientific companion is physiology. Indeed you may wonder, as we proceed, whether behaviorism can be differentiated from that science. It is different from physiology only in the grouping of its problems, not in fundamentals or in central viewpoint. Physiology is particularly interested in the functioning of parts of the animal—for example, its digestive system, the circulatory system, the nervous system, the excretory systems, the mechanics of neural and muscular response. Behaviorism, on the other hand, while it is intensely interested in all of the functioning of these parts, is intrinsically interested in what the whole animal will do from morning to night and from night to morning.

The interest of the behaviorist in man's doings is more than the interest of the spectator—he wants to control man's reactions as physical scientists want to control and manipulate other natural phenomena. It is the business of behavioristic psychology to be able to predict and to control human activity. To do this it must gather scientific data by experimental methods. Only then can the trained behaviorist predict, given the stimulus, what reaction will take place; or, given the reaction, state what the situation or stimulus is that has caused the reaction.

B. F. Skinner

Taking up the behaviorist standard from Watson, the Harvard professor of psychology B. F. Skinner repeats the oft-made suggestion to apply the methods of science to human affairs. Like many before him, he takes it as obvious that a causal explanation of human behavior must exclude all concepts of free will, apparently unaware of arguments to the contrary in the empiricist tradition (represented in the excerpts from Hume in this anthology). Skinner is more unusual in his blaming of much of "the present unhappy condition of the world" on our alleged confusion between a "scientific conception of human behavior" and a "philosophy of personal freedom." At any rate, Skinner thinks he has the answer to practical as well as theoretical problems in his theory of behavior.

Skinner's commitment to Watson's behaviorist program is so rigorous that he believes, and argues here, that all reference to inner events, whether physiological or mental, must be extruded not only from the *data,* but from the explanatory *theories* of a scientific psychology. (More recent psychology has not followed this methodological purism.) Thus he leads us into his animal laboratory to study the "external variables" of which, he believes, behavior is a function. But notice how, having introduced his terminology of "reinforcement" to describe experiments with pigeons, Skinner applies it in the very next paragraph to complex human actions such as playing games, flying planes, and talking. The assumption that is concealed under this quick transition is that the vocabulary and theories adequate for describing and explaining the simplest elements of animal movement are equally appropriate for all kinds of intentional performance. This is questioned by K. Lorenz (even for much animal behavior) and by N. Chomsky (especially for human language).

Skinner obviously thinks that certain principles governing what he calls "reinforcement" can explain all human learning, including the acquiring of one's mother tongue, as well as every influence from one's society. His enthusiasm for social engineering would be congenial to Plato, and perhaps to Marx, but his conception of it ("making the human product meet more acceptable specifications") is more redolent of the capitalist factory. These excerpts are from *Science and Human Behavior* (Macmillan, 1953), Chapters I, III, V, XXVII, XXVIII, and XXIX.

THE APPLICATION OF SCIENCE TO MAN,
AND THE THREAT TO ORDINARY NOTIONS OF FREEDOM

The methods of science have been enormously successful wherever they have been tried. Let us then apply them to human affairs. We need not retreat in those sectors where science has already advanced. It is necessary only to bring our understanding of human nature up to the same point. Indeed this may well be our only hope. If we can observe human behavior carefully from an objective point of view and come to understand it for what it is, we may be able to adopt a more sensible course of action. The need for establishing some such balance is now widely felt, and those who are able to control the direction of science are acting accordingly. It is understood that there is no point in furthering a science of nature unless it includes a sizable science of human nature, because only in that case will the results be wisely used. It is possible that science has come to the rescue and that order will eventually be achieved in the field of human affairs.

There is one difficulty, however. The application of science to human behavior is not so simple as it seems. Most of those who advocate it are simply looking for "the facts." To them science is little more than careful observation. They want to evaluate human behavior as it really is rather than as it appears to be through ignorance or prejudice, and then to make effective decisions and move on rapidly to a happier world. But the way in which science has been applied in other fields shows that something more is involved. Science is not concerned just with "getting the facts," after which one may act with greater wisdom in an unscientific fashion. Science supplies its own wisdom. It leads to a new conception of a subject matter, a new way of thinking about that part of the world to which it has addressed itself. If we are to enjoy the advantages of science in the field of human affairs, we must be prepared to adopt the working model of behavior to which a science will inevitably lead. But very few of those who advocate the application of scientific method to current problems are willing to go that far.

Science is more than the mere description of events as they occur. It is an attempt to discover order, to show that certain events stand in lawful relations to other events. No practical technology can be based upon science until such relations have been discovered. But order is not only a possible end product; it is a working assumption which must be adopted at the very start. We cannot apply the methods of science to a subject matter which is assumed to move about capriciously. Science not only describes, it predicts. It deals not only with the past but with the future. Nor is prediction the last word: to the extent that relevant conditions can be altered, or otherwise controlled, the future can be controlled. If we are to use the methods of science in the field of human affairs, we must assume that behavior is lawful and determined. We must expect to discover that what a man does is the result of specifiable conditions and that once these conditions have been discovered, we can anticipate and to some extent determine his actions.

This possibility is offensive to many people. It is opposed to a tradition of long standing which regards man as a free agent, whose behavior is the product, not of specifiable antecedent conditions, but of spontaneous inner changes of course. Prevailing philosophies of human nature recognize an internal "will" which has the power of interfering with causal relationships and which makes the prediction and control of behavior impossible. To suggest that we abandon this view is to threaten many cherished beliefs—to undermine what appears to be a stimulating and productive conception of human nature. The alternative point of view insists upon recognizing coercive forces in human conduct which we may prefer to disregard. It challenges our aspirations, either worldly or otherworldly. Regardless of how much we stand to gain from supposing that human behavior is the proper subject matter of a science, no one who is a product of Western civilization can do so without a struggle. We simply do not want such a science.

Conflicts of this sort are not unknown in the history of science. When Aesop's lion was shown a painting in which a man was depicted killing a lion, he commented contemptuously, "The artist was obviously a man." Primitive beliefs about man and his place in nature are usually flattering. It has been the unfortunate responsibility of science to paint more realistic pictures. The Copernican theory of the solar system displaced man from his pre-eminent position at the center of things. Today we accept this theory without emotion, but originally it met with enormous resistance. Darwin challenged a practice of segregation in which man set himself firmly apart from the animals, and the bitter struggle which arose is not yet ended. But though Darwin put man in his biological place, he did not deny him a possible position as master. Special faculties or a special capacity for spontaneous, creative action might have emerged in the process of evolution. When that distinction is now questioned, a new threat arises.

There are many ways of hedging on the theoretical issue. It may be insisted that a science of human behavior is impossible, that behavior has certain essential features which forever keep it beyond the pale of science. But although this argument may dissuade many people from further inquiry, it is not likely to have any effect upon those who are willing to try and see. Another objection frequently offered is that science is appropriate up to a certain point, but that there must always remain an area in which one can act only on faith or with respect to a "value judgment": science may tell us *how* to deal with human behavior, but just *what* is to be done must be decided in an essentially nonscientific way. Or it may be argued that there is another kind of science which is compatible with doctrines of personal freedom. For example, the social sciences are sometimes said to be fundamentally different from the natural sciences and not concerned with the same kinds of lawfulness. Prediction and control may be forsworn in favor of "interpretation" or some other species of understanding. But the kinds of intellectual activities exemplified by value judgments or by intuition or interpretation have never been set forth clearly, nor have they yet shown any capacity to work a change in our present predicament.

Our current practices do not represent any well-defined theoretical position. They are, in fact, thoroughly confused. At times we appear to regard a man's behavior as spontaneous and responsible. At other times we recognize that inner determination is at least not complete, that the individual is not always to be held to account.

We have not been able to reject the slowly accumulating evidence that circumstances beyond the individual are relevant. We sometimes exonerate a man by pointing to "extenuating circumstances." We no longer blame the uneducated for their ignorance or call the unemployed lazy. We no longer hold children wholly accountable for their delinquencies. "Ignorance of the law" is no longer wholly inexcusable: "Father, forgive them; for they know not what they do." The insane have long since been cleared of responsibility for their condition, and the kinds of neurotic or psychotic behavior to which we now apply this extenuation are multiplying.

But we have not gone all the way. We regard the common man as the product of his environment; yet we reserve the right to give personal credit to great men for their achievements. (At the same time we take a certain delight in proving that part of the output of even such men is due to the "influence" of other men or to some trivial circumstance in their personal history.) We want to believe that right-minded men are moved by valid principles even though we are willing to regard wrong-minded men as victims of erroneous propaganda. Backward peoples may be the fault of a poor culture, but we want to regard the elite as something more than the product of a good culture. Though we observe that Moslem children in general become Moslems while Christian children in general become Christians, we are not willing to accept an accident of birth as a basis for belief. We dismiss those who disagree with us as victims of ignorance, but we regard the promotion of our own religious beliefs as something more than the arrangement of a particular environment.

All of this suggests that we are in transition. We have not wholly abandoned the traditional philosophy of human nature; at the same time we are far from adopting a scientific point of view without reservation. We have accepted the assumption of determinism in part; yet we allow our sympathies, our first allegiances, and our personal aspirations to rise to the defense of the traditional view. We are currently engaged in a sort of patchwork in which new facts and methods are assembled in accordance with traditional theories.

If this were a theoretical issue only, we would have no cause for alarm; but theories affect practices. A scientific conception of human behavior dictates one practice, a philosophy of personal freedom another. Confusion in theory means confusion in practice. The present unhappy condition of the world may in large measure be traced to our vacillation. The principal issues in dispute between nations, both in peaceful assembly and on the battlefield, are intimately concerned with the problem of human freedom and control. Totalitarianism or democracy, the state of the individual, planned society or laissez-faire, the impression of cultures upon alien peoples, economic determinism, individual initiative, propaganda, education, ideological warfare—all concern the fundamental nature of human behavior. We shall almost certainly remain ineffective in solving these problems until we adopt a consistent point of view.

We cannot really evaluate the issue until we understand the alternatives. The traditional view of human nature in Western culture is well known. The conception of a free, responsible individual is embedded in our language and pervades our practices, codes, and beliefs. Given an example of human behavior, most people can describe it immediately in terms of such a conception. The practice is so natural that it is seldom examined. A scientific formulation, on the other hand, is new and strange. Very few people have any notion of the extent to which a science of human behavior is indeed possible. In what way can the behavior of the individual or of groups of individuals be predicted and controlled? What are laws of behavior like? What overall conception of the human organism as a behaving system emerges? It is only when we have answered these questions, at least in a preliminary fashion, that we may consider the implications of a science of human behavior with respect to either a theory of human nature or the management of human affairs. . . .

THE EXTERNAL CAUSES OF BEHAVIOR

We are concerned, then, with the causes of human behavior. We want to know why men behave as they do. Any condition or event which can be shown to have an effect upon behavior must be taken into account. By discovering and analyzing these causes we can predict behavior; to the extent that we can manipulate them, we can control behavior. . . .

Eventually a science of the nervous system based upon direct observation rather than inference will describe the neural states and events which immediately precede instances of behavior. We shall know the precise neurological conditions which immediately precede, say, the response, "No, thank you." These events in turn will be found to be preceded by other neurological events, and these in turn by others. This series will lead us back to events outside the nervous system and, eventually, outside the organism. . . . [W]e shall consider external events of this sort in some detail. We shall then be better able to evaluate the place of neurological explanations of behavior. However, we may note here that we do not have and may never have this sort of neurological information at the moment it is needed in order to predict a specific instance of behavior. It is even more unlikely that we shall be able to alter the nervous system directly in order to set up the antecedent conditions of a particular instance. The causes to be sought in the nervous system are, therefore, of limited usefulness in the prediction and control of specific behavior.

Psychic inner causes.

An even more common practice is to explain behavior in terms of an inner agent which lacks physical dimensions and is called "mental" or "psychic." The purest form of the psychic explanation is seen in the animism of primitive peoples. From the immobility of the body after death it is inferred that a spirit responsible for movement has departed. The *enthusiastic* person is, as the etymology of the word implies, energized by a "god within." It is only a modest refinement to attribute every feature of the behavior of the physical organism to a corresponding feature of the "mind" or of some inner "personality." The inner man is regarded as driving the body very much as the man at the steering wheel drives a car. The inner man wills an action, the other executes it. The inner loses his appetite, the outer stops eating. The inner man wants and the outer gets. The inner has the impulse which the outer obeys.

It is not the layman alone who resorts to these practices, for many reputable psychologists use a similar dualistic system of explanation. The inner man is sometimes personified clearly, as when delinquent behavior is attributed to a "disordered personality," or he may be dealt with in fragments, as when behavior is attributed to mental processes, faculties, and traits. Since the inner man does not occupy space, he may be multiplied at will. It has been argued that a single physical organism is controlled by several psychic agents and that its behavior is the resultant of their several wills. The Freudian concepts of the ego, superego, and id are often used in this way. They are frequently regarded as nonsubstantial creatures, often in violent conflict, whose defeats or victories lead to the adjusted or maladjusted behavior of the physical organism in which they reside.

Direct observation of the mind comparable with the observation of the nervous system has not proved feasible. It is true that many people believe that they observe their "mental states" just as the physiologist observes neural events, but another interpretation of what they observe is possible. . . . Introspective psychology no longer pretends to supply direct information about events which are the causal antecedents, rather than the mere accompaniments, of behavior. It defines its "subjective" events in ways which strip them of any usefulness in a causal analysis. The events appealed to in early mentalistic explanations of behavior have remained beyond the reach of observation. Freud insisted upon this by emphasizing the role of the unconscious—a frank recognition that important mental processes are not directly observable. The Freudian literature supplies many examples of behavior from which unconscious wishes, impulses, instincts, and emotions are inferred. Unconscious thought-processes have also been used to explain intellectual achievements. Though the mathematician may feel that he knows "how he thinks," he is often unable to give a coherent account of the mental processes leading to the solution of a specific problem. But any mental event which is unconscious is necessarily inferential, and the explanation is therefore not based upon independent observations of a valid cause. . . .

The practice of looking inside the organism for an explanation of behavior has tended to obscure the variables which are immediately available for a scientific analysis. These variables lie outside the organism, in its immediate environment and in its environmental history. They have a physical status to which the usual techniques of science are adapted, and they make it possible to explain behavior as other subjects are explained in science. These independent variables are of many sorts and their relations to behavior are often subtle and complex, but we cannot hope to give an adequate account of behavior without analyzing them.

Consider the act of drinking a glass of water. . . . To what extent is it helpful to be told, "He drinks because he is thirsty"? If to be thirsty means nothing more than to have a tendency to drink, this is mere redundancy. If it means that he drinks because of a state of thirst, an inner causal event is invoked. If this state is purely inferential—if no dimensions are assigned to it which would make direct observation possible—it cannot serve as an explanation. But if it has physiological or psychic properties, what role can it play in a science of behavior?

The physiologist may point out that several ways of raising the probability of drinking have a common effect: they increase the concentration of solutions in the body. Through some mechanism not yet well understood, this may bring about a corresponding change in the nervous system which in turn makes drinking more probable. In the same way, it may be argued that all these operations make the organism "feel thirsty" or "want a drink" and that such a psychic state also acts upon the nervous system in some unexplained way to induce drinking. In each case we have a causal chain consisting of three links: (1) an operation performed upon the organism from without—for example, water deprivation; (2) an inner condition—for example, physiological or psychic thirst; and (3) a kind of behavior—for example, drinking. Independent information about the second link would obviously permit us to predict the third without recourse to the first. It would be a preferred type of variable because it would be nonhistoric; the first link may lie in the past history of the organism, but the second is a current condition. Direct information about the second link is, however, seldom, if ever, available. Sometimes we infer the second link from the third: an animal is judged to be thirsty if it drinks. In that case, the explanation is spurious. Sometimes we infer the second link from the first: an animal is said to be thirsty if it has not drunk for a long time. In that case, we obviously cannot dispense with the prior history.

The second link is useless in the *control* of behavior unless we can manipulate it. At the moment, we have no way of directly altering neural processes at appropriate moments in the life of a behaving organism, nor has any way been discovered to alter a psychic process. We usually set up the second link through the first: we make an animal thirsty, in either the physiological or the psychic sense, by depriving it of water, feeding it salt, and so on. In that case, the second link obviously does not permit us to dispense with the first. Even if some new technical discovery were to enable us to set up or change the second link directly, we should still have to deal with those enormous areas in which human behavior is controlled through manipulation of the first link. A technique of operating upon the second link would increase our control of behavior, but the techniques which have already been developed would still remain to be analyzed.

The most objectionable practice is to follow the causal sequence back only as far as a hypothetical second link. This is a serious handicap both in a theoretical science and in the practical control of behavior. It is no help to be told that to get an organism to drink we are simply to "make it thirsty" unless we are also told how this is to be done. When we have obtained the necessary prescription for thirst, the whole proposal is more complex than it need be. Similarly, when an example of maladjusted behavior is explained by saying that the individual is "suffering from anxiety," we have still to be told the cause of the anxiety. But the external conditions which are then invoked could have been directly related to the maladjusted behavior. Again, when we are told that a man stole a loaf of bread because "he was hungry," we have still to learn of the external conditions responsible for the "hunger." These conditions would have sufficed to explain the theft.

The objection to inner states is not that they do not exist, but that they are not relevant in a functional analysis. We cannot account for the behavior of any system while staying wholly inside it; eventually we must turn to forces operating upon the organism from without. Unless there is a weak spot in our causal chain so that the second link is not lawfully determined by the first, or the third by the second, then the first and third links must be lawfully related. If we must always go back beyond the second link for prediction and control, we may avoid many tiresome and exhausting digressions by examining the third link as a function of the first. Valid information about the second link may throw light upon this relationship but can in no way alter it.

The external variables of which behavior is a function provide for what may be called a causal or functional analysis. We undertake to predict and control the behavior of the individual organism. This is our "dependent variable"—the effect for which we are to find the cause. Our "independent variables"—the causes of behavior—are the external conditions of which behavior is a function. Relations between the two—the "cause-and-effect relationships" in behavior—are the laws of a science. A synthesis of these laws expressed in quantitative terms yields a comprehensive picture of the organism as a behaving system.

This must be done within the bounds of a natural science. We cannot assume that behavior has any peculiar properties which require methods or special kinds of knowledge. It is often argued that an act is not so important as the "intent" which lies behind it, or that it can be described only in terms of what it "means" to the behaving individual or to others whom it may affect. If statements of this sort are useful for scientific purposes, they must be based upon observable events, and we may confine ourselves to such events exclusively in a functional analysis. We shall see later that although such terms as "meaning" and "intent" appear to refer to properties of behavior, they usually conceal references to independent variables. This is also true of "aggressive," "friendly," "disorganized," "intelligent," and other terms which appear to describe properties of behavior but in reality refer to its controlling relations.

The independent variables must also be described in physical terms. An effort is often made to avoid the labor of analyzing a physical situation by guessing what it "means" to an organism or by distinguishing between the physical world and psychological world of "experience." This practice also reflects a confusion between dependent and independent variables. The events affecting an organism must be capable of description in the language of physical science. It is sometimes argued that certain "social forces" or the "influences" of culture or tradition are exceptions. But we cannot appeal to entities of this sort without explaining how they can affect both the scientist and the individual under observation. The physical events which must then be appealed to in such an explanation will supply us with alternative material suitable for a physical analysis.

By confining ourselves to these observable events, we gain a considerable advantage, not only in theory, but in practice. A "social force" is no more useful in manipulating behavior than an inner state of hunger, anxiety, or skepticism. Just as we must trace these inner events to the manipulable variables of which they are said to be functions before we may put them to practical use, so we must identify the physical events through which a "social force" is said to affect the organism before we can manipulate it for purposes of control. In dealing with the directly observable data we need not refer to either the inner state or the outer force. . . .

THE EXPERIMENTAL STUDY OF CONDITIONING
AND REINFORCEMENT

We select a relatively simple bit of behavior which may be freely and rapidly repeated, and which is easily observed and recorded. If our experimental subject is a pigeon, for example, the behavior of raising the head above a given height is convenient. This may be observed by sighting across the pigeon's head at a scale pinned on the far wall of the box. We first study the height at which the head is normally held and select some line on the scale which is reached only infrequently. Keeping our eye on the scale we then begin to open the food tray very quickly whenever the head rises above the line. If the experiment is conducted according to specifications, the result is invariable: we observe an immediate change in the frequency with which the head crosses the line. . . .

The statement that the bird "learns that it will get food by stretching its neck" is an inaccurate report of what has happened. To say that it has acquired the "habit" of stretching its neck is merely to resort to an explanatory fiction, since our only evidence of the habit is the acquired tendency to perform the act. The barest

possible statement of the process is this: we make a given consequence contingent upon certain physical properties of behavior (the
upward movement of the head), and the behavior is then observed
to increase in frequency.

It is customary to refer to any movement of the organism as a
"response." The word is borrowed from the field of reflex action
and implies an act which, so to speak, answers a prior event—the
stimulus. But we may make an event contingent upon behavior
without identifying, or being able to identify, a prior stimulus. We
did not alter the environment of the pigeon to *elicit* the upward
movement of the head. It is probably impossible to show that any
single stimulus invariably precedes this movement. Behavior of this
sort may come under the control of stimuli, but the relation is not
that of elicitation. The term "response" is therefore not wholly
appropriate but is so well established that we shall use it in the
following discussion.

A response which has already occurred cannot, of course, be
predicted or controlled. We can only predict that *similar* responses
will occur in the future. The unit of a predictive science is, therefore, not a response but a class of responses. The word "operant"
will be used to describe this class. The term emphasizes the fact that
the behavior sequences define the properties with respect to which
responses are called similar. The term will be used both as an adjec-
tive (operant behavior) and as a noun to designate the behavior
defined by a given consequence. . . .

The term "learning" may profitably be saved in its traditional
sense to describe the reassortment of responses in a complex situation. Terms for the process of stamping in may be borrowed from
Pavlov's analysis of the conditioned reflex. Pavlov himself called all
events which strengthened behavior "reinforcement" and all the
resulting changes "conditioning." In the Pavlovian experiment,
however, a reinforcer is paired with a *stimulus;* whereas in operant
behavior it is contingent upon a *response.* Operant reinforcement is
therefore a separate process and requires a separate analysis. In both
cases, the strengthening of behavior which results from reinforcement is appropriately called "conditioning." In operant conditioning we "strengthen" an operant in the sense of making a response
more probable or, in actual fact, more frequent. In Pavlovian or

"respondent" conditioning we simply increase the magnitude of the response elicited by the conditioned stimulus and shorten the time which elapses between stimulus and response. (We note, incidentally, that these two cases exhaust the possibilities: an organism is conditioned when a reinforcer [1] accompanies another stimulus or [2] follows upon the organism's own behavior. Any event which does neither has no effect in changing a probability of response.) In the pigeon experiment, then, food is the *reinforcer* and presenting food when a response is emitted is the *reinforcement*. The *operant* is defined by the property upon which reinforcement is contingent— the height to which the head must be raised. The change in frequency with which the head is lifted to this height is the process of *operant conditioning.*

While we are awake, we act upon the environment constantly, and many of the consequences of our actions are reinforcing. Through operant conditioning the environment builds the basic repertoire with which we keep our balance, walk, play games, handle instruments and tools, talk, write, sail a boat, drive a car, or fly a plane. A change in the environment—a new car, a new friend, a new field of interest, a new job, a new location—may find us unprepared, but our behavior usually adjusts quickly as we acquire new responses and discard old. We shall see in the following chapter that operant reinforcement does more than build a behavioral repertoire. It improves the efficiency of behavior and maintains behavior in strength long after acquisition or efficiency has ceased to be of interest. . . .

In dealing with our fellow men in everyday life and in the clinic and laboratory, we may need to know just how reinforcing a specific event is. We often begin by noting the extent to which our own behavior is reinforced by the same event. This practice frequently miscarries; yet it is still commonly believed that reinforcers can be identified apart from their effects upon a particular organism. As the term is used here, however, the only defining characteristic of a reinforcing stimulus is that it reinforces.

The only way to tell whether or not a given event is reinforcing to a given organism under given conditions is to make a direct test. We observe the frequency of a selected response, then make an event contingent upon it and observe any change in frequency. If there is a change, we classify the event as reinforcing to the organism under the existing conditions. There is nothing circular about classifying events in terms of their effects; the criterion is both empirical and objective. It would be circular, however, if we then went on to assert that a given event strengthens an operant *because* it is reinforcing. We achieve a certain success in guessing at reinforcing powers only because we have in a sense made a crude survey; we have gauged the reinforcing effect of a stimulus upon ourselves and assume the same effect upon others. We are successful only when we resemble the organism under study and when we have correctly surveyed our own behavior.

Events which are found to be reinforcing are of two sorts. Some reinforcements consist of *presenting* stimuli, of adding something—for example, food, water, or sexual contact—to the situation. These we call *positive* reinforcers. Others consist of *removing* something—for example, a loud noise, a very bright light, extreme cold or heat, or electric shock—from the situation. These we call *negative* reinforcers. In both cases the effect of reinforcement is the same—the probability of response is increased. We cannot avoid this distinction by arguing that what is reinforcing in the negative case is the *absence* of the bright light, loud noise, and so on; for it is absence after presence which is effective, and this is only another way of saying that the stimulus is removed. The difference between the two cases will be clearer when we consider the *presentation* of a *negative* reinforcer or the *removal* of a *positive*. These are the consequences which we call punishment.

A survey of the events which reinforce a given individual is often required in the practical application of operant conditioning. In every field in which human behavior figures prominently—education, government, the family, the clinic, industry, art, literature, and so on—we are constantly changing probabilities of response by arranging reinforcing consequences. The industrialist who wants employees to work consistently and without absenteeism must make certain that their behavior is suitably reinforced—not only with wages but with suitable working conditions. The girl who wants another date must be sure that her friend's behavior in inviting her and in keeping the appointment is suitably reinforced. To teach a child to read or sing or play a game effectively, we must work out a program of educational reinforcement in which appropriate responses "pay off" frequently. If the patient is to return for further counsel, the psychotherapist must make sure that the behavior of coming to him is in some measure reinforced. . . .

THE INFLUENCE OF CULTURE,
AND THE POSSIBILITY OF CHANGING IT

[T]he individual acquires from the group an extensive repertoire of *manners* and *customs*. What a man eats and drinks and how he does so, what sorts of sexual behavior he engages in, how he builds a house or draws a picture or rows a boat, what subjects he talks about or remains silent about, what music he makes, what kinds of personal relationships he enters into and what kinds he avoids—all depend in part upon the practices of the group of which he is a member. The actual manners and customs of many groups have, of course, been extensively described by sociologists and anthropologists. Here we are concerned only with the kinds of processes which they exemplify.

Behavior comes to conform to the standards of a given community when certain responses are reinforced and others are allowed to go unreinforced or are punished. These consequences are often closely interwoven with those of the nonsocial environment. The way in which a man rows a boat, for example, depends in part upon certain mechanical contingencies; some movements are effective and others ineffective in propelling the boat. These contingencies depend upon the construction of the boat and oars—which are in turn the result of other practices observed by the boatmakers in the group. They also depend upon the type of water, which may be peculiar to a group for geographical reasons, so that the manner in which a boat is rowed in an inland lake district is different from that along the seacoast even when boat and oars are of the same type. The educational contingencies established by the group are still another source of difference. The individual is reinforced with approval when he adopts certain grips, postures, kinds of strokes, and so on, and punished with criticism when he adopts others. These variables are especially important in determining the "style" which eventually becomes characteristic of a group.

The contingencies to be observed in the social environment easily explain the behavior of the conforming individual. The problem is to explain the contingencies. Some of these are arranged for reasons which have no connection with the effect of customs or manners upon the group. The community functions as a reinforcing environment in which certain kinds of behavior are reinforced and others punished, but it is maintained as such through other return benefits. Verbal behavior is a good example. In a given community certain vocal responses are characteristically reinforced with food, water, and other services or objects. These responses become part of a child's repertoire as naturally as nonverbal responses reinforced by the same consequences. It does not greatly matter whether a child gets a drink by bending over a pool or by saying, "I want a drink of water." To explain why the water is forthcoming in the latter case, however, requires a rather elaborate analysis of the verbal environment. It is enough to note here that a verbal environment may maintain itself through its effects upon all participants, quite apart from its function in teaching the language to new members of the community. An adult in a new verbal environment may receive no explicit educational reinforcement but may nevertheless acquire an adequate vocabulary. Some nonverbal customs and manners can be explained in the same way. Moreover, when a custom is perpetuated by a governmental, religious, or educational agency, we may point to the usual return benefits. . . .

Perhaps the simplest explanation of the differential reinforcement of conforming behavior is the process of induction. The forces which shape ethical behavior to group standards are powerful. The group steps in to suppress lying, stealing, physical assault, and so on, because of immediate consequences to its members. Its behavior in so doing is eventually a function of certain characteristic features of the "good" and "bad" behavior of the controlled individual. Among these is lack of conformity to the general behavior of the group. There is thus a frequent association of aversive properties of behavior with the property of nonconformance to a standard. Nonconforming behavior is not always aversive, but aversive behavior is always nonconforming. If these properties are paired often enough, the property of nonconformance becomes aversive. "Right" and "wrong" eventually have the force of "conforming" and "nonconforming." Instances of behavior which are nonconforming but not otherwise aversive to the group are henceforth treated as if they were aversive.

No matter how we ultimately explain the action of the group in extending the ethical classification of "right" and "wrong" to manners and customs, we are on solid ground in observing the contingencies by virtue of which the behavior characteristic of a particular group is maintained. As each individual comes to conform to a standard pattern of conduct, he also comes to support that pattern by applying a similar classification to the behavior of others. Moreover, his own conforming behavior contributes to the standard with which the behavior of others is compared. Once a custom, manner, or style has arisen, therefore, the social system which observes it appears to be reasonably self-sustaining.

A social environment is usually spoken of as the "culture" of a group. The term is often supposed to refer to a spirit or atmosphere or something with equally nonphysical dimensions. Our analysis of the social environment, however, provides an account of the essential features of culture within the framework of a natural science. It permits us not only to understand the effect of culture but, as we shall see later, to alter cultural design.

In the broadest possible sense the culture into which an individual is born is composed of all the variables affecting him which are arranged by other people. The social environment is in part the result of those practices of the group which generate ethical behavior and of the extension of these practices to manners and customs. It is in part the accomplishment of all the agencies, . . . and various subagencies with which the individual may be in especially close contact. The individual's family, for example, may control him through an extension of religious or governmental techniques, by way of psychotherapy, through economic control, or as an educational institution. The special groups to which he belongs—from the play group or street gang to adult social organizations—have similar effects. Particular individuals may also exert special forms of control. A culture, in this broad sense, is thus enormously complex and extraordinarily powerful. . . .

The social environment of any group of people is the product of a complex series of events in which accident sometimes plays a prominent role. Manners and customs often spring from circumstances which have little or no relation to the ultimate effect upon the group. The origins of more explicit controlling practices may be equally adventitious. Thus the pattern of control exercised by a strong leader, reflecting many of his personal idiosyncrasies, may result in an established governmental classification of behavior as legal or illegal and may even set the pattern for a highly organized agency. The techniques which a saint employs to control himself may become part of the established practices of a religious agency. Economic control is determined in part by the resources available to the group, which are ultimately a matter of geography. Other fortuitous factors are introduced when different cultures intermingle or when a culture survives important changes in the nonsocial environment. A cultural practice is not the less effective in determining the behavior characteristic of a group because its origins are accidental. But once the effect upon behavior has been observed, the source of the practice may be scrutinized more closely. Certain questions come to be asked. Why should the design of a culture be left so largely to accident? Is it not possible to change the social environment deliberately so that the human product will meet more acceptable specifications? . . .

. . . [T]he most effective control from the point of view of survival will probably be based upon the most reliable estimates of the survival value of cultural practices. Since a science of behavior is concerned with demonstrating the consequences of cultural practices we have some reason for believing that such a science will be an essential mark of the culture or cultures which survive. The current culture which, on this score alone, is most likely to survive is, therefore, that in which the methods of science are most effectively applied to the problems of human behavior.

This does not mean, however, that scientists are becoming self-appointed governors. It does not mean that anyone in possession of the methods and results of science can step outside the stream of history and take the evolution of government into his own hands. Science is not free, either. It cannot interfere with the course of events; it is simply part of that course. It would be quite inconsistent if we were to exempt the scientist from the account which science gives of human behavior in general. Science can, however, supply a description of the kind of process of which it itself is an example. A reasonable statement of our present position in the evolution of culture might take this form: We find ourselves members of a culture in which science has flourished and in which the methods of science have come to be applied to human behavior. If, as seems to be the case, the culture derives strength from this fact, it is a reasonable prediction that a science of behavior will continue to flourish and that our culture will make a substantial contribution to the social environment of the future.

Konrad Lorenz

The Austrian biologist Konrad Lorenz recently won a Nobel Prize for his work in founding the new discipline of ethology. The ethologists accepted the behaviorists' definition of their common subject of study—the behavior of living creatures—but they disagreed sharply about the kind of explanation of behavior to be expected and about the most appropriate methods of research. Whereas the behaviorists hoped to find some laws of learning, relating behavior to environment (response to stimulus), which would apply to *all* animals, including humans, the ethologists backed the more plausible hunch that there are innate behavioral differences between species. They, therefore, concentrated much more on observing what different animals do *spontaneously,* before intervening to experiment, and were prepared to explain much animal behavior in terms of instinctual tendencies, whose presence in the species is in turn to be explained by evolutionary pressures.

Like Skinner, Lorenz has always been ready and eager to apply his scientific theories to diagnose human ills. He has published a number of popular works, of which the most important is *On Aggression* (1963), but these extracts are taken from an earlier paper, "Part and Parcel in Animal and Human Societies," translated by R. Martin in *Studies in Animal and Human Behaviour,* Vol. II (Harvard University Press, 1971). The reader must judge for himself whether Lorenz's inferences from animals to humans are any more plausible than Skinner's.

THE ETHOLOGICAL APPROACH
TO THE EXPLANATION OF BEHAVIOR

This article is directed against the widely-occurring methodological error found in sociology and ethnological psychology, which consists in utterly neglecting the fixed structural properties of the human individual which are *not* accessible to influence exerted by the totality of the society. This methodological error is derived from two sources, which are considered separately.

The first source—*false generalization of principles of Gestalt psychology*—is discussed in the first section [not reprinted here].

The second source of the methodological error laid bare in this article rests in neglect of *innate, species-specific behaviour patterns*. . . .

There follows, in compressed form, the research history relating to the nature and peculiarity of the two most important individually invariable components of animal and human behaviour—the *endogenous—automatic motor* and the *innate releasing mechanism*. . . .

The first great discovery, which can more-or-less be described as the birth of comparative behavioural research was doubtless the identification of a genuine phylogenetic *homology* between innate, species-specific motor patterns of related animal species. C. O. Whitman and O. Heinroth were both highly trained and extremely knowledgeable comparative morphologists, and by way of a lucky—but far from chance—coincidence they also possessed the extremely broad mass knowledge of behaviour patterns of a group of related animals which is the prerequisite for phylogenetic comparison. Thus it was that they first made—independently of one another—the simple, but extremely fruitful, discovery that there are motor patterns of constant form which are performed in exactly the same manner by every healthy individual of a species. These represent characteristic features not only for the individual species, but (just like many "conservative" morphological characters) also for even the broadest groups of related forms. In other words, they discovered motor patterns whose distribution and occurrence in the animal kingdom intimately resembles that of *organs*. One can thus safely assume that the developmental history of these patterns during speciation has followed the same lines as that of organs and organ characteristics. Therefore, their phylogenetic age can be determined with the same methodological means as that of morphological characters, and it follows that the concept of phyletic homology can be applied to the former in just the same way as to the latter. Both researchers demonstrated this fact by employing suitable characters of such motor patterns as taxonomic characters in the reconstruction of phylogenetic relationships within a given group of related animals. The end-product was then compared with conclusions obtained from the fine systematic evaluation of morphological characters. The unqualified agreement between the results convincingly demonstrated the correctness of their initial belief, which had been summarized by Whitman in 1898: "Instincts and organs are to be studied from the common viewpoint of phyletic descent."

The newly discovered facts demonstrated equally convincingly the utter untenability of the ideas which had been previously developed with respect to the nature of so-called "instinctive" behaviour, both on the vitalistic-teleological side and on the mechanistic side. In the view of the vitalists, "the instinct" was a "directive factor" fundamentally immune to causal explanation, and the behaviour determined by instinct was necessarily characterized by the typical, plastic variability of all *purposive* behaviour. Thus, it was actually quite logical when representatives of "purposive psychology" quite simply equated the species-preserving function of species-specific innate behaviour patterns with a purpose towards which the animal subject strives. On the mechanistic side, by contrast, there were—as already mentioned—*two* schools of thought. That of the behaviourists stated that complex, innate behaviour patterns are just as much figments of the imagination as innate "goals" and that the species-preserving appropriate form of so-called instinctive behaviour is only *apparently innate*—in reality it is only acquired in the individual lifetime of every organism through trial and error (Watson). The Pavlovian reflexologist school admitted the existence of highly differentiated and lengthy innate motor sequences, but explained them as chain assemblies of unconditioned reflexes. Quite early on, the correct objection was raised by *purposive psychologists,* above all McDougall, that the *spontaneity* of many "instinctive" behaviour patterns and their obvious great independence of external stimuli cannot be explained through the reflex principle. . . .

The main fact which struck home was an unexpected correlation between spontaneity and individual invariability of species-specific motor patterns. According to the vitalistic-teleological view, a spontaneous, goal-directed behaviour pattern aimed at a specific "drive goal" must *ipso facto* have a constant biological end-result and yet exhibit variability in the motor co-ordinations involved. According to the reflex-chain theory, on the other hand, a chain of reflexly co-ordinated individual movements determined in a rigidly innate fashion throughout its entire length cannot exhibit spontaneity. But in reality it was *just these* behaviour patterns with an overall innate, species-specific character which exhibited a particular, extremely peculiar tendency towards spontaneous eruption independent of external stimuli. The innate releasing mechanism, which in most cases acts as a "response lock" ensuring selective response of innate, species-specific behaviour patterns in specific environmental situations, does indeed prove to be reliably reactive, but the motor pattern itself exhibits extremely peculiar behaviour which cannot at all be explained through the reflex principle. . . . Nowadays, we know from the investigations of E. von Holst, P. Weiss and others that innate, species-specific motor patterns are not based on conditioned and unconditioned reflexes, like so many other behaviour patterns of animals and man. *They are based instead upon a different elementary function of the central nervous system,* namely upon a spontaneous, automatic-rhythmic *production* of stimuli. . . .

INNATE FEATURES IN HUMAN SOCIAL BEHAVIOR

Without a trace of doubt, human beings exhibit the *smallest* range of endogenous-automatic motor patterns found in any higher organism. Apart from certain motor norms of food-uptake (seizing, placing-in-the-mouth, chewing and swallowing), mating (frictional movements) and possibly certain automatic elements in walking and running, an adult human being appears to have virtually no centrally co-ordinated motor patterns based on endogenous automatisms. However, this paucity of genuine instinctive motor patterns is not a primary feature but the result of a *process of reduction.* In the case of most instinctive motor patterns of man, only *motor display patterns* have remained. Where these owe their origin to intention movements, they permit recognition of the original form of the motor pattern—as was seen with the display of anger which was correctly interpreted by Darwin as a formalized intention movement.

We are, in principle, in agreement with McDougall's view that the qualitatively separate emotions of human beings each correspond to an "instinct" in the sense of an action-specific arousal modality and behavioural motivation. With very many of these specific behavioural motivations, assumption of an endogenous-automatic basis is supported by the fact that the stimulus threshold sinks during periods without response and rises following abreaction of the drive concerned. This applies to coarse sexual responses and also to the completely independent behaviour patterns of falling-in-love, to demonstrative behaviour and to other patterns. Endogenous accumulation of response-specific behavioural motivation is most clearly evident, however, with the behaviour patterns which were interpreted by Freud as products of the *aggressive drive*. Anyone who has ever worked under a mildly excitable and not quite self-controlled supervisor knows that the emergence of a "threatening atmosphere" is a quite periodic phenomenon and that, after eruption and abreaction of the dammed-up arousal in a "purifying thunderstorm," the good will of the despot is not reduced but definitely augmented. After a "normal office row," a peculiar atmosphere of increased human affection prevails! One of my aunts regularly became involved in arguments with her house maid, each time giving notice to leave. The typical shift in the perceptual field which, as is well known, accompanies the changes in the existing cumulative value of response-specific energy, expressed itself wonderfully in my aunt's behaviour in that she was inordinately pleased with each replacement encountered immediately after the discharge of her aggressive condition. She never realized that these "pearls" were compulsively and repeatedly converted to more-or-less hateful creatures in the course of a few months. The accumulation of aggressive responses can become very disruptive, even dangerous, when a very small community is completely isolated from conspecific surroundings on which the dammed-up drives can be abreacted. The "polar malady" which emerges among members of expeditions, the crew of small boats, and so on, is nothing less than massive threshold-lowering of the behaviour patterns of angry outbreaks. Anybody who has become acquainted with this knows how minor irritations can eventually have a ridiculously angering effect. Even when one has full insight into one's own response, one is unable to prevent oneself from boiling with rage at certain small characteristics of a companion, for example, a cough or a peculiar way of speaking. In so responding to a friend, one behaves fundamentally just like the male of an isolated Cichlid fish pair which will finally attack and kill the female following the absence of threatening conspecifics to chase away from the family. This response pattern is particularly typical of *Geophagus,* and one can prevent this by placing a mirror in the aquarium for the male to abreact his aggression.

Another response pattern, which will be given special treatment here because of its particular importance, is that of *social defence.* Its subjective experiential correlate is the emotion of *enthusiasm.* An especially interesting feature of the response is its association with motor patterns *which are homologous with those of the chimpanzee.* Any man with strong emotions knows from his own experience the overpowering feelings of the onlooker in moments when combative involvement for the society is elicited in us. This emotion is evoked by erection of the hairs on the neck, the upper part of the back and—interestingly—on the outside surface of the upper arm. The gestural motor display patterns consist of pronounced erectness of the body posture, raising of the head, furrowing of the brows, downward retraction of the corners of the mouth, forward projection of the lower jaw, forward extension of the shoulders, and rotation of the arms on the shoulder joints such that the hairy dorsal surface is pointed outwards. The facial expression is the same as that which evokes in us the described emotional responses to the "dummy" of the eagle's head. Inward rotation of the arms and the contraction of the *musculi arrectores pilorum* does not contribute much to the visual impression of the overall behaviour in human beings, and it would presumably not have been noticed if anthropoid apes did not possess exactly the same motor patterns. A chimpanzee advancing to carry out social defence similarly pushes out his chin, rotates his arms inwards and erects the hair on the upper back and the outside of the upper arm. (On these areas of the body, the hairs have undergone elongation to enhance the releaser concerned, just as is the case with very many pelage and plumage areas of mammals and birds adapted for an analogous function.) As a result of the forward-inclined body posture and the relatively great length of the arms in the ape, these motor patterns produce a considerable enlargement of the body outline, which has a demonstrative, intimidatory function and quite definitely has an "infectious" effect on fellow members of the society. It is a neat example of a *rudimentary* behaviour pattern, in the true phylogenetic sense, that human beings in analogous cases "spread out the hair which they no longer have"! The response is elicited by a very simple relationship schema; one which one can justifiably describe as "unfortunate," since—however valuable it may be for internal cohesion of societies—the innate incorrigibility of its response and particularly the nature of the releasing situation carry great inherent dangers to humanity. The vital relationship character of the releasing mechanism concerned is an external *threat* to the society, and demagogues through the ages have misused this (in itself ethically valuable) response to set nations at one another's throats with the simple dummy of a fabricated enemy and a fabricated threat to the society. . . .

THE EVOLUTIONARY EXPLANATION OF PROBLEMS IN HUMAN NATURE

Arnold Gehlen's comment that man is a *"cultural being by nature"* proves to be a bold conception which is convincingly correct in more than one aspect, from the standpoint of comparative behavioural research. It has already been stated that the human being represents the "instinct-reduction organism." The "openness to the environment" exhibited by human beings, which is regarded as a constitutive character by Gehlen—the extensive freedom from specific, hereditarily-determined adaptations to the environment—is a feature largely resulting from domestication-induced deficiencies in fixed, innate action and response norms. It was already recognized by Charles Otis Whitman, one of the first pioneers of comparative behavioural research, that loss of "instincts" in domestic animals by no means signifies a retrogressive step in the development of intelligence—as would be expected from the theory of Spencer and Lloyd Morgan, who held "the instinct" to be the phylogenetic predecessor of intelligent behaviour. Whitman demonstrated with a series of excellent observational examples that domesticated animals are frequently capable of solving by insight problems which the wild-form cannot master, and that the exclusive reason for this is that they have attained novel degrees of freedom in plastically pur-posive behaviour by loss of certain fixed, instinctive action and response norms to which the non-domesticated form of the same animal species remains firmly bound. In reference to the relation-ship between domestication-induced instinct deficiencies and the insight function which these render possible, Whitman made the notable statement: "These 'faults of instincts' are not intelligence, but they are the open door, through which the great educator expe-rience comes in and performs every miracle of intellect."

Whitman himself established an initial link between the specific "openness to the environment" and behavioural liberty of human beings on the one hand, and such domestication-induced deficien-cies on the other. It is quite definite that the disappearance or ex-pansion of certain innate releasing mechanisms is an indispensable prerequisite for the versatility and cosmopolitan nature of man. Our closest phylogenetic relatives, the anthropoids, are all special-ized for narrowly defined ecological niches and the (in geological terms) abrupt transition from the very "stenecic" ancestral form to the extreme "euryecic" mode of life of human beings cannot be explained on the basis of the normal processes of speciation. This can only be understood through the assumption of domestication-induced dedifferentiation processes taking place in fixed innate mechanisms, which would be possible within a few centuries. . . .

Gehlen recognizes a further constitutive aspect of man in that the human being represents the endangered or "jeopardized" creature—the creature "with a constitutional chance to encounter mishaps." From what has been said about domestication phenomena as a pre-requisite of humanization, it is sufficiently obvious that this concep-tion presented by Gehlen also incorporates a profound truth. Origin of the specifically human characteristic of behavioural liberty was quite definitely dependent upon the reduction and disruption of rigidly structured action and response norms. Like all other rigid structures, innate behaviour patterns also exhibit the property of supporting and reinforcing. The consequent, adaptable "rigidity" of behaviour could only be dismantled at the expense of the sup-porting function and therefore with the loss of safeguards. Each new element of behavioural plasticity had to be bought in exchange for certain degrees of security.

Every process of organic further development—particularly with respect to intelligence—is always a peculiar compromise between these two inseparable and yet so contrasting aspects of all rigid structures. Without rigid structures, no organic system with a high level of integration is possible; but in every case the structures of the existing system must be undone if a system with a still higher level of integration and harmony is to be attained. This unfortunate dilemma is a fundamental attribute of all organic further developments. When a crustacean moults, when a human being undergoes an alteration in personality structure from that of the child to that of a man at puberty, or when an archaic human social order is transformed to a new one, the developmental advance is in every case accompanied by dangers. The reason for this is that the old structure must be dismantled before the new one has achieved full functional capacity. No other organism is or was exposed to these dangers to the same extent as man, since no other in the entire history of life on our planet has experienced, or is experiencing, such headlong development. Phylogenetically and ontogenetically, man is "the incomplete creature"; ontogenetically and phylogenetically, he is involved in a virtually continuous series of "moults," and is never in the state of statically equilibriated structural adaptation which can persist for geological epochs of enormous duration with other organisms.

There are very few philosophical dicta which maintain the opposite of the truth so fundamentally as the old saying: *natura non facit saltum!* From the events within the atom to those in the history of mankind, inorganic and organic developments occur in leaps. Even though certain cases of quantitative summation in developmental processes may appear to be continuous when superficially observed, they are basically just as discontinuous as the large qualitative changes occurring in organic development, which were first, clearly recognized as such by Hegel. The dangers discussed above as inherent in all of the more significant leaps in organic developmental processes are, understandably, proportional to the magnitude of the individual leaps. We are therefore not at all surprised that one of the greatest qualitative changes that has ever taken place in the history of organic events—that of the evolutionary leap from the anthropoid ape to man, which (in geological terms) took place so extremely rapidly—is accompanied by enormous dangers for the newly evolved creature.

Even somebody who doubts the hypotheses of comparative behavioural research sketched above—in particular the role played by domestication—must admit that the structures whose dismantling has led to the endangerment of mankind are those of *innate behaviour*. The price which man had to pay for his constitutive liberty in thinking processes and behavioural acts is that of adaptedness to a specific environment and a specific form of social existence, which in all sub-human organisms is guaranteed by species-specific, inherited action and response norms. This adaptedness signifies, in the social context, nothing other than complete agreement between inclination and necessity, i.e. the problemless life in paradise which had to be sacrificed in return for the fruits of the tree of knowledge.

The functional process which substitutes for the lost "instincts" in man is that of dialogue-like, *investigatory* interaction with the environment, the process of "coming to an understanding" with external reality as is etymologically expressed in the word *Vernunft (reason)*. Man is the creature of reason. But he is not *exclusively* a creature of reason; his behaviour is by no means so thoroughly determined by reason as is assumed by the majority of philosophical anthropologists. There is a far greater degree of influence exerted by innate, species-specific action and response norms that we usually like to believe and accept. This particularly applies to human *social* behaviour. It has been stated above that in animals of advanced intelligence behaviour towards a conspecific is controlled far more by innate components and far less by higher intelligent functions than behaviour towards the extra-specific environment. That this is unfortunately equally true of human beings as well is crudely expressed in the incongruity between man's mastery of his external environment and his stunning inability to solve the intraspecific problems of mankind.

This is by no means due to the fact that these intraspecific (in the broadest sense, *social*) problems are in any way more difficult than those of the external environment. The opposite is true. The fission of the atom, without a trace of doubt, presents the human mind with far more difficult tasks than the question of how one can prevent human beings from wiping one another out with the aid of atom bombs. There are many people of above-average intelligence whose ability for abstract thought is simply not adequate for performing the non-descriptive, mathematical thought processes on which modern atomic physics is based. On the other hand, even somebody of sub-normal intelligence can immediately see what would have to happen, and what must be avoided, in order to prevent the self-destruction of mankind. Despite the enormous difference in conceptual difficulties between these two problems, mankind has solved that of atomic processes within a few decades, whilst man is today even more helpless against the danger of self-destruction—which emerged with the invention of the first weapon, the hand-axe—than at the time of Pekin man!

The fact that even the most modest powers of reason are adequate for recognition of what should not happen, and that this nevertheless occurs, gives much food for thought. Wherever something of this nature occurs in the more transparent behaviour of the human individual, where self-observation is possible (i.e. where, despite full insight into a situation of vital importance, something contrary to reason nevertheless irresistibly takes its course), there is almost always *blockage* of the application of rational thought by dominant innate, species-specific action and response patterns. It is a fairly obvious step to assume the same for analogous, supra-individual, collective failure of rational application. At least, this assumption is promising enough as a working hypothesis for disfunction of innate species-specific behaviour patterns to be seriously taken into account as a probable cause of the otherwise completely unintelligible failure of collective human reason in the face of relatively simple problems. I shall therefore give a compressed account of a series of possible—indeed, probable—disruptive mechanisms of social behaviour patterns.

THE PROBLEM OF HUMAN AGGRESSION

The first, and probably most important, of these mechanisms is the following: With every organism that is plucked out of its natural environment and placed in novel surroundings, behaviour patterns occur which are neutral or even detrimental for the survival of the species. This phenomenon always arises through the fact that a specific behaviour pattern, based upon endogenous stimulus-production and adapted for a quite specific, species-preserving function, is robbed of its normal releasing situation, such that accumulation of action-specific energy leads to its eruption in a completely inadequate stimulus situation. This is referred to as a *misplaced response*.

Modern man represents such an animal, torn from his natural environmental niche. Within a space of time which—from the geological-phylogenetic standpoint—is immeasurably short, the flowering of human culture has so extensively changed the entire ecology and sociology of our species that a whole range of previously adaptive endogenous behaviour patterns have become not only non-functional but extremely disruptive. We have become acquainted with the most important example of this in the so-called aggressive drive. For a chimpanzee, and even for early stone-age man, it would be doubtless quite valuable and necessary in the sense of maintenance of individuals, families and the species that internal stimulus-production should be adequate to produce, say, two major outbursts of rage per week. But since such innate action/response norms, in so far as they are not affected by sudden deficiency mutations induced by domestication, can only change at the same phylogenetic tempo as organic structures, it is not particularly surprising that modern man, immersed in his police-protected existence, does not know what to do with these rhythmically-occurring outbursts of rage. In colloquial language, there is a very neat expression of the human quality of "looking for a good row," summarizing the fact that a relatively harmless discharge of this accumulated response-specific energy has the value of curative catharsis. The increased motivation for aggression, which is a result of just this accumulation of response-specific energy, is doubtless the cause of the easily *incitable* nature of man. Everybody is pleased to some degree to find a "permissible" substitute object for his stored aggressive tendencies and happily seizes upon the clumsiest "dummy" offered by a clever demagogue. I maintain that without this purely physiological basis all of the past cases of demagogically directed mass cruelty, such as witchcraft trials or anti-semitic persecution, would have been fundamentally impossible.

A very similar role of "misplaced response" is played by the more complex behaviour pattern of social defence, which has also been mentioned previously. As already indicated, this response is also only too easily evoked by dummies presented by demagogues. Because of the highly pleasurable character of its subjective experiential correlate, that of social or national fervour, this response is particularly dangerous. It should be unreservedly admitted that it is a magnificent experience to sing the national anthem, overcome by a "holy" paroxysm, and it is only too easy to forget that the paroxysm represents the hair-erection of the old chimpanzee pelage and that the entire response is fundamentally directed *against* one "enemy" or another. Above all, nowadays when cave-bears and sabre-toothed tigers no longer exist as dangers to human society, this "enemy" is always a society of fellow human beings, who feel just as committed to enthusiastic defence of *their* society! The social and (in the profoundest sense of the word) ethical value which doubtless exists in the *unifying* function of the human response concerned will only be accessible to mankind when we have learnt to insert into the "schema" of the enemy the dangers which really threaten humanity and not a chosen group of fellow human beings set up by a demagogue!

The next mechanism of disruption affecting social behaviour patterns to be discussed finds its cause, just like the preceding case, in the headlong change in social conditions, with which the phylogenetic mutability of species-specific innate action/response norms cannot keep pace. But in this case the disruption is not based on the fact that *lesser* exhaustion of action-specific stimulus-production provides a disruptive excess of impulses; instead, rapid further differentiation of the human social order has led to the necessity for a *greater occurence* of certain social behaviour patterns which our innate system of social action and response patterns is unable to provide. Among the culturally determined alterations of human social life involved, the predominant effect is probably based on the change from an originally *closed* form of human society to an *anonymous* type. With a very large number of innate social response patterns of human beings, one of the prerequisites for release is that the fellow human being towards whom they are directed should be a personally acquainted member of the society—a "friend." Among the ten commandments, those which concern behaviour towards fellow men are followed as a matter of course from natural inclination without any demand upon responsible, moral decision, as long as the "neighbour" is a personally well-acquainted friend and companion. When one remembers what was said in the section covering animal behaviour patterns exhibiting moral analogies, and calls to mind the close-knit nature of the society represented by a chimpanzee troop—despite jealousy and rank-order squabbles, then one must assume that innate social behaviour patterns played at least the same role in primitive human groups. We fundamentally disagree with the biblical thesis defended by some child psychologists and psychoanalysts, that man is utterly "wicked from childhood on." Our primitive ancestor, before the actual process of "humanization," was—at the very least—just as "good" as a wolf or chimpanzee, in both of which youngsters and females are spared from harm and in which even a jealously fought male society member is protected against an external enemy without hesitation and with all possible vigour. I should like to paraphrase the biblical text to read: Man is not wicked from childhood on, but just good enough for the requirements which were made of him in the primitive human group, in which the small number of individuals were personally acquainted with one another and "loved" one another after their own fashion. He is not good *enough* for the requirements set by the enormously expanded, anonymous society of later cultural epochs, which demand that he should behave towards any completely unacquainted fellow human being as if he were a personal friend.

Another, extremely special case of this described phenomenon of increasing inaccessibility of innate social action/response norms also entered into the penetrating, abrupt change in human relationships brought about by the *invention of weapons*. One must call to mind what was said in the section on behavioural systems exhibiting ethic analogies, where we discussed the finely-balanced equilibrium which operates between the ability to kill and innate killing inhibitions of an animal species. It then becomes immediately clear how great a danger is presented for the further survival of the species by any disruption of this equilibrium favouring the side of killing ability. Whenever an old male chimpanzee, despite his powerful aggressive drives, refrains from killing or even damaging a weaker conspecific (which would be to the detriment of species survival), it is thanks to species-specific innate inhibitions, which are elicited by highly specific releasing mechanisms. We have become acquainted with "submissive" postures and calls which represent releasers of such inhibitions. These "sympathy-releasing" inhibition-evoking factors are, of course, only operative for the species-specific, somewhat slow and gruesome methods of killing with the natural weapons of the species. If one now imagines that such an excitable and vicious creature is quite suddenly presented with a "more humane" method of killing, whose extremely rapid operation completely excludes functioning of the described inhibition-releasers, then one can understand the terrible consequences which the invention of weapons—from the hand-axe to the atom-bomb—had and has for mankind. The biological situation is fundamentaly the same as if a gruesome freak of nature had suddenly equipped the turtle-dove, which (as we have seen) possesses no killing inhibitions whatsoever and no equivalent releasing mechanisms, with the beak of a raven, but without at the same time fitting the dove with the inhibitory mechanisms which are correlated with the presence of this weapon in the raven. However, it must immediately be added, as a corollary to this, that in exact terms the invention of a weapon by a *non-responsible* animal is impossible; the example is purely imaginary and simply intended to demonstrate the mechanism of endangerment. The emergence of a genuine invention, such as that of the hand-axe, is dependent upon an extremely high degree of differentiation of dialogue-like, investigative interaction with the environment, which itself is very close to genuine setting of questions and understanding of the answers. The ability for invention and that for responsibility are dependent upon the same prerequisites.

The failure of killing inhibitions in the situation produced by the invention of weapons is principally based on the fact that the innate mechanisms releasing an inhibition do not respond in the new situation. The deeper, emotional layers of our natures to some extent fail to "understand" the consequences of the use of weapons, and it is obviously not quite sufficient that reason should grasp something which is emotionally inaccessible. The amazingly small degree of inhibition which thoroughly sympathetic and agreeable civilized human beings exhibit with respect to killing animals, or even other human beings in battle, is entirely understandable on this basis of inability to "understand." Like all other emotions, those of sympathy only respond to situations for which we have a predetermined, innate receptor correlate. If an armed man were brought to realize the actual consequences of his actions—the fact that through tensing of his finger on a trigger he can rip out the entrails of a living creature—in a manner not only accessible to his reason, but also immediately accessible to his emotions (i.e. in a manner *similar to that operating in the use of natural weapons*), only very few human beings would hunt for pleasure and the vast majority would be conscientious objectors. The firing of a long-range gun or the dropping of a bomb is so thoroughly "impersonal" that normal human beings who are absolutely incapable of throttling a mortal enemy with their hands are nevertheless able, without further ado, to apply finger-pressure to deliver up thousands of women and children to a horrible death. . . .

THE FUNCTIONAL LIMITATIONS OF RATIONAL MORALITY

The aim of this article is to demonstrate the absolute obligatory requirement for consideration of innate, species-specific action/response norms of human beings in the investigation of inter-human relationships. It is not intended to discuss the nature of rational thought and its specific compensatory and regulatory function in human beings. . . . Nobody is further from underestimating the difference between man and animals than the comparative behavioural investigator, and nobody can more clearly calculate the absolute *novelty* of the enormous regulative powers over maintenance and further development of existence which are inherent in the rational responsibility of man. The character of a hitherto unknown, novel creation which is exhibited by our moral law affects the comparative behavioural investigator more than any other "with ever-recurring feelings of bewonderment." I maintain that one can only recognize the uniqueness of man in its entire, impressive magnitude when this may be *separated* from the background of old, historic characteristics which man still shares today with the higher animals.

But what we are concerned with in this article is just exactly this old aspect of human behaviour, which can only be explained historically and which in fact does *not* obey the laws of rational, responsible morality. The misinterpretation against which this article is directed is the general *overestimation* of the influence exerted by the totality of human society on the structure of the individual and the combined *underestimation* of the influence exerted on the structure and function of the supra-individual society by fixed, phyletically/historically transmitted structures of the individual. As regards theories of morality, this misinterpretation is represented in overestimation of the function of rational morality and underestimation of the role played even in human beings by moral-analogous systems of innate behaviour, such as we have encountered in animal behaviour. I continue to hold the opinion expressed in a previous publication, ". . . that in no single case of regularly occurring action by an individual human being bearing upon the welfare and woe of the society is there exclusive operation of the categorical imperative providing impulse and motive for the altruistic act. Instead, in the great majority of cases, the primary active impulse is provided by the response of innate schemata and inherited drives. It is exceedingly difficult to construct situations which are really indifferent as realeasers of innate responses, but at the same time demand of us an active and altruistic attitude along the pathway of rational consideration."

On the other hand, I do not believe that I am guilty of the equivalent misinterpretation of *underestimating* the function of responsible thought in social behaviour of modern man. The culturally determined changes in human ecology and sociology are so profound that hardly one of our natural "inclinations" is *completely* sufficient to meet the requirements of modern social structure. A supplementary spur in the form of a categorical "thou shalt" or a supplementary inhibition in the form of a "thou shalt not" is necessary at every stage. No modern human being can give free rein to his innate inclinations, and presumably this provides the basis for the age-old human longing for the "lost paradise." Not only civilized men, but all human beings everywhere are afflicted by "discomfort in their culture" (S. Freud). No human being is "happy" in the sense applicable to a wild animal, whose innate inclinations are entirely in accord with what "should be done" in the interests of species-survival. . . .

. . . I am fully conscious of the fact that all of what can be said from comparative behavioural research about the role played by functional disruption of innate, species-specific behaviour patterns in the catastrophic social confusion of humanity must provisionally remain purely hypothetical in nature. All the same, I see the most important *practical* task of our branch of research in profound investigation of these hypotheses. One only has to think of the almost overwhelming importance which would accrue in the pedagogical, corrective, educational and (above all) ethnological-psychological realms to an initial discovery of *which* disruptions in social behaviour are influenced by rational morality (and therefore by education) and which are not. What we do know is that both phenomena exist. One must remember that the dangers threatening modern man with extinction are derived exclusively from disruptions in social behaviour; it is not the external environment, but man himself who is the threat to humanity. It is only necessary to recall the horrifying simile of the turtle-dove equipped by a gruesome freak of nature with the beak of a raven. Nobody can deny that man finds himself in an analogous position.

One thing, however, is not simply hypothetical but established truth: *The only means of eliminating a functional disruption in a system lies in causal analysis of the system and the disruption.* One might possibly be able to "understand" an archway through intuitive, holistic appreciation, without possessing knowledge of the form and function of the component building-blocks, but one is unable to *repair* it. Even the queen of applied science—medicine—owes her ability to re-establish a functional entity, which has gone awry, entirely to causal analysis of the component functions. At the moment, humanity represents a functional entity which has gone fundamentally awry. The misunderstanding between the development of the weapon and the inhibitions against using it threatens to extinguish mankind. Will it be possible for us, through collective human striving for knowledge and the collective responsibility of all human beings, to restore the equilibrium between ability to exterminate and social inhibition, which (just like so many other reliable, static conditions of equilibrium) had to be sacrificed in exchange for the dynamic developmental opportunity for human thought and behaviour? The future of mankind will be decided by the answering of this question!

Noam Chomsky

The American linguist Noam Chomsky has become famous for setting the study of language on a new course in the 1960s. In his technical works he has presented his evolving theories of "transformational grammar," seeking general principles which underlie the incredible complexity of human languages. Chomsky has increasingly related his linguistic work to psychology and philosophy—indeed, he has done much to erase the borderlines between these disciplines. His review of Skinner's book *Verbal Behavior* (in *Language,* Vol. 35, 1959) is widely regarded as a conclusive demolition of behaviorist accounts of language learning. He has defended what he takes to be the insights of the rationalist tradition in philosophy, arguing that a substantial element in what everyone knows—for instance, about the limits to the grammatical forms that any human language can take—is innate in us, not learned from experience.

Whether or not Chomsky is right in the details of his theories about innate factors in language, his main methodological lesson is surely vital for all the social sciences; that is, before we even begin to ask questions about how a certain knowledge or skill or practice is acquired by individuals, developed in a culture, or evolved as innate in a species, we must first make sure we have an adequate description of what exactly it is about which we are asking these questions. That this prior conceptual task is by no means trivial or easy is shown by the radically impoverished conception of human language that the behaviorists tended to assume without question.

These extracts are from the lectures entitled "Language and Mind," delivered at the University of California at Berkeley (first published by Harcourt, Brace and World, Inc., 1968).

THE INADEQUACY OF BEHAVIORIST ACCOUNTS
OF LANGUAGE

In these lectures, I would like to focus attention on the question, What contribution can the study of language make to our understanding of human nature? In one or another manifestation, this question threads its way through modern Western thought. In an age that was less self-conscious and less compartmentalized than ours, the nature of language, the respects in which language mirrors human mental processes or shapes the flow and character of thought—these were topics for study and speculation by scholars and gifted amateurs with a wide variety of interests, points of view, and intellectual backgrounds. And in the nineteenth and twentieth centuries, as linguistics, philosophy, and psychology have uneasily tried to go their separate ways, the classical problems of language and mind have inevitably reappeared and have served to link these diverging fields and to give direction and significance to their efforts. There have been signs in the past decade that the rather artificial separation of disciplines may be coming to an end. It is no longer a point of honor for each to demonstrate its absolute independence of the others, and new interests have emerged that permit the classical problems to be formulated in novel and occasionally suggestive ways—for example, in terms of the new perspectives provided by cybernetics and the communication sciences, and against the background of developments in comparative and physiological psychology that challenge long-standing convictions and free the scientific imagination from certain shackles that had become so familiar a part of our intellectual environment as to be almost beyond awareness. All of this is highly encouraging. I think there is more of a healthy ferment in cognitive psychology—and in the particular branch of cognitive psychology known as linguistics—than there has been for many years. And one of the most encouraging signs is that skepticism with regard to the orthodoxies of the recent past is coupled with an awareness of temptations and the dangers of premature orthodoxy, an awareness that, if it can persist, may prevent the rise of new and stultifying dogma.

It is easy to be misled in an assessment of the current scene; nevertheless, it seems to me that the decline of dogmatism and the accompanying search for new approaches to old and often still intractable problems are quite unmistakable, not only in linguistics but in all of the disciplines concerned with the study of mind. I remember quite clearly my own feeling of uneasiness as a student at the fact that, so it seemed, the basic problems of the field were solved, and that what remained was to sharpen and improve techniques of linguistic analysis that were reasonably well understood and to apply them to a wider range of linguistic materials. In the postwar years, this was a dominant attitude in most active centers of research. I recall being told by a distinguished anthropological linguist, in 1953, that he had no intention of working through a vast collection of materials that he had assembled because within a few years it would surely be possible to program a computer to construct a grammar from a large corpus of data by the use of techniques that were already fairly well formalized. At the time, this did not seem an unreasonable attitude, though the prospect was saddening for anyone who felt, or at least hoped, that the resources of human intelligence were somewhat deeper than these procedures and techniques might reveal. Correspondingly, there was a striking decline in studies of linguistic method in the early 1950's as the most active theoretical minds turned to the problem of how an essentially closed body of technique could be applied to some new domain—say, to analysis of connected discourse, or to other cultural phenomena beyond language. I arrived at Harvard as a graduate student shortly after B. F. Skinner had delivered his William James Lectures, later to be published in his book *Verbal Behavior*. Among those active in research in the philosophy or psychology of language, there was then little doubt that although details were missing, and although matters could not really be quite that simple, nevertheless a behavioristic framework of the sort Skinner had outlined would prove quite adequate to accommodate the full range of language use. There was now little reason to question the conviction of Leonard Bloomfield, Bertrand Russell, and positivistic linguists, psychologists, and philosophers in general that the framework of stimulus-response [S-R] psychology would soon be extended to the point where it would provide a satisfying explanation for the most mysterious of human abilities. The most radical souls felt that perhaps, in order to do full justice to these abilities, one must postulate little *s*'s and *r*'s *inside the brain alongside the capital S's and R's* that were open to immediate inspection, but this extension was not inconsistent with the general picture.

Critical voices, even those that commanded considerable prestige, were simply unheard. For example, Karl Lashley gave a brilliant critique of the prevailing framework of ideas in 1948, arguing that underlying language use—and all organized behavior—there must be abstract mechanisms of some sort that are not analyzable in terms of association and that could not have been developed by any such simple means. But his arguments and proposals, though sound and perceptive, had absolutely no effect on the development of the field and went by unnoticed even at his own university (Harvard), then the leading center of psycholinguistic research. Ten years later Lashley's contribution began to be appreciated, but only after his insights had been independently achieved in another context.

The technological advances of the 1940's simply reinforced the general euphoria. Computers were on the horizon, and their imminent availability reinforced the belief that it would suffice to gain a theoretical understanding of only the simplest and most superficially obvious of phenomena—everything else would merely prove to be "more of the same," an apparent complexity that would be disentangled by the electronic marvels. The sound spectrograph, developed during the war, offered similar promise for the physical analysis of speech sounds. The interdisciplinary conferences on speech analysis of the early 1950's make interesting reading today. There were few so benighted as to question the possibility, in fact the immediacy, of a final solution to the problem of converting speech into writing by available engineering technique. And just a few years later, it was jubilantly discovered that machine translation and automatic abstracting were also just around the corner. For those who sought a more mathematical formulation of the basic processes, there was the newly developed mathematical theory of communication, which, it was widely believed in the early 1950's, had provided a fundamental concept—the concept of "information"—that would unify the social and behavioral sciences and permit the development of a solid and satisfactory mathematical theory of human behavior on a probabilistic base. At about the same time, the theory of automata developed as an independent study, making use of closely related mathematical notions. And it was linked at once, and quite properly, to earlier explorations of the theory of neural nets. There were those—John von Neumann, for example—who felt that the entire development was dubious and shaky at best, and probably quite misconceived, but such qualms did not go far to dispel the feeling that mathematics, technology, and behavioristic linguistics and psychology were converging on a point of view that was very simple, very clear, and fully adequate to provide a basic understanding of what tradition had left shrouded in mystery.

In the United States at least, there is little trace today of the illusions of the early postwar years. If we consider the current status of structural linguistic methodology, stimulus-response psycholinguistics (whether or not extended to "mediation theory"), or probabilistic or automata-theoretic models for language use, we find that in each case a parallel development has taken place: A careful analysis has shown that insofar as the system of concepts and principles that was advanced can be made precise, it can be demonstrated to be inadequate in a fundamental way. The kinds of structures that are realizable in terms of these theories are simply not those that must be postulated to underlie the use of language, if empirical conditions of adequacy are to be satisfied. What is more, the character of the failure and inadequacy is such as to give little reason to believe that these approaches are on the right track. That is, in each case it has been argued—quite persuasively, in my opinion—that the approach is not only inadequate but misguided in basic and important ways. It has, I believe, become quite clear that if we are ever to understand how language is used or acquired, then we must abstract for separate and independent study a cognitive system, a system of knowledge and belief, that develops in early childhood and that interacts with many other factors to determine the kinds of behavior that we observe; to introduce a technical term we must isolate and study the system of *linguistic competence* that underlies behavior but that is not realized in any direct or simple way in behavior. And this system of linguistic competence is qualitatively different from anything that can be described in terms of the taxonomic methods of structural linguistics, the concepts of S-R psychology, or the notions developed within the mathematical theory of communication or the theory of simple automata. The theories and models that were developed to describe simple and immediately given phenomena cannot incorporate the real system of linguistic competence; "extrapolation" for simple descriptions cannot approach the reality of linguistic competence; mental structures are not simply "more of the same" but are qualitatively different from the complex networks and structures that can be developed by elaboration of the concepts that seemed so promising to many scientists just a few years ago. What is involved is not a matter of degree of complexity but rather a quality of complexity. Correspondingly, there is no reason to expect that the available technology can provide significant insight or understanding or useful achievements; it has noticeably failed to do so, and, in fact, an appreciable investment of time, energy, and money in the use of computers for linguistic research—appreciable by the standards of a small field like linguistics—has not provided any significant advance in our understanding of the use or nature of language. These judgments are harsh, but I think they are defensible. They are, furthermore, hardly debated by active linguistic or psycholinguistic researchers.

At the same time there have been significant advances, I believe, in our understanding of the nature of linguistic competence and some of the ways in which it is put to use, but these advances, such as they are, have proceeded from assumptions very different from those that were so enthusiastically put forth in the period I have been discussing. What is more, these advances have not narrowed the gap between what is known and what can be seen to lie beyond the scope of present understanding and technique; rather, each advance has made it clear that these intellectual horizons are far more remote than was heretofore imagined. Finally, it has become fairly clear, it seems to me, that the assumptions and approaches that appear to be productive today have a distinctly traditional flavor to them; in general, a much despised tradition has been largely revitalized in recent years and its contributions given some serious and, I believe, well-deserved attention. From the recognition of these facts flows the general and quite healthy attitude of skepticism that I spoke of earlier.

In short, it seems to me quite appropriate, at this moment in the development of linguistics and psychology in general, to turn again to classical questions and to ask what new insights have been achieved that bear on them, and how the classical issues may provide direction for contemporary research and study.

When we turn to the history of study and speculation concerning the nature of mind and, more specifically, the nature of human language, our attention quite naturally comes to focus on the seventeenth century, "the century of genius," in which the foundations of modern science were firmly established and the problems that still confound us were formulated with remarkable clarity and perspicuity. There are many far from superficial respects in which the intellectual climate of today resembles that of seventeenth-century Western Europe. One, particularly crucial in the present context, is the very great interest in the potentialities and capacities of automata, a problem that intrigued the seventeenth-century mind as fully as it does our own. I mentioned above that there is a slowly dawning realization that a significant gap—more accurately, a yawning chasm—separates the system of concepts of which we have a fairly clear grasp, on the one hand, and the nature of human intelligence, on the other. A similar realization lies at the base of Cartesian philosophy. Descartes also arrived, quite early in his investigations, at the conclusion that the study of mind faces us with a problem of quality of complexity, not merely degree of complexity. He felt that he had demonstrated that understanding and will, the two fundamental properties of the human mind, involved capacities and principles that are not realizable by even the most complex of automata.

It is particularly interesting to trace the development of this argument in the works of the minor and now quite forgotten Cartesian philosophers, like Cordemoy, who wrote a fascinating treatise extending Descartes' few remarks about language, or La Forge, who produced a long and detailed *Traité de l'esprit de l'homme* expressing, so he claimed with some reason, what Descartes would likely have said about this subject had he lived to extend his theory of man beyond physiology. One may question the details of this argument, and one can show how it was impeded and distorted by certain remnants of scholastic doctrine—the framework of substance and mode, for example. But the general structure of the argument is not unreasonable; it is, in fact, rather analogous to the argument against the framework of ideas of the early postwar years, which I mentioned at the outset of this lecture. The Cartesians tried to show that when the theory of corporeal body is sharpened and clarified and extended to its limits, it is still incapable of accounting for facts that are obvious to introspection and that are also confirmed by our observation of the actions of other humans. In particular, it cannot account for the normal use of human language, just as it cannot explain the basic properties of thought. Consequently, it becomes necessary to invoke an entirely new principle—in Cartesian terms, to postulate a second substance whose essence is thought, alongside of body, with its essential properties of extension and motion. This new principle has a "creative aspect," which is evidenced most clearly in what we may refer to as "the creative aspect of language use," the distinctively human ability to express new thoughts and to understand entirely new expressions of thought, within the framework of an "instituted language," a language that is a cultural product subject to laws and principles partially unique to it and partially reflections of general properties of mind. These laws and principles, it is maintained, are not formulable in terms of even the most elaborate extension of concepts proper to the analysis of behavior and interaction of physical bodies, and they are not realizable by even the most complex automaton. In fact, Descartes argued that the only sure indication that another body possesses a human mind, that it is not a mere automaton, is its ability to use language in the normal way; and he argued that this ability cannot be detected in an animal or an automaton which, in other respects, shows signs of apparent intelligence exceeding those of a human, even though such an organism or machine might be as fully endowed as a human with the physiological organs necessary to produce speech.

I will return to this argument and the ways in which it was developed. But I think it is important to stress that, with all its gaps and deficiencies, it is an argument that must be taken seriously. There is nothing at all absurd in the conclusion. It seems to me quite possible that at that particular moment in the development of Western thought there was the possibility for the birth of a science of psychology of a sort that still does not exist, a psychology that begins with the problem of characterizing various systems of human knowledge and belief, the concepts in terms of which they are organized and the principles that underlie them, and that only then turns to the study of how these systems might have developed through some combination of innate structure and organism-environment interaction. Such a psychology would contrast rather sharply with the approach to human intelligence that begins by postulating, on a priori grounds, certain specific mechanisms that, it is claimed, *must* be those underlying the acquisition of all knowledge and belief. The distinction is one to which I will return in a subsequent lecture. For the moment, I want merely to stress the reasonableness of the rejected alternative and, what is more, its consistency with the approach that proved so successful in the seventeenth-century revolution in physics. . . .

The theory of generative grammar, both particular and universal, points to a conceptual lacuna in psychological theory that I believe is worth mentioning. Psychology conceived as "behavioral science" has been concerned with behavior and acquisition or control of behavior. It has no concept corresponding to "competence," in the sense in which competence is characterized by a generative grammar. The theory of learning has limited itself to a narrow and surely inadequate concept of what is learned—namely a system of stimulus-response connections, a network of associations, a repertoire of behavioral items, a habit hierarchy, or a system of dispositions to respond in a particular way under specifiable stimulus conditions. Insofar as behavioral psychology has been applied to education or therapy, it has correspondingly limited itself to this concept of "what is learned." But a generative grammar cannot be characterized in these terms. What is necessary, in addition to the concept of behavior and learning, is a concept of what is learned—a notion of competence—that lies beyond the conceptual limits of behaviorist psychological theory. Like much of modern linguistics and modern philosophy of language, behaviorist psychology has quite consciously accepted methodological restrictions that do not permit study of systems of the necessary complexity and abstractness. One important future contribution of the study of language to general psychology may be to focus attention on this conceptual gap and to demonstrate how it may be filled by the elaboration of a system of underlying competence in one domain of human intelligence.

There is an obvious sense in which any aspect of psychology is based ultimately on the observation of behavior. But it is not at all obvious that the study of learning should proceed directly to the investigation of factors that control behavior or of conditions under which a "behavioral repertoire" is established. It is first necessary to determine the significant characteristics of this behavioral repertoire, the principles on which it is organized. A meaningful study of learning can proceed only after this preliminary task has been carried out and has led to a reasonably well-confirmed theory of underlying competence—in the case of language, to use the formulation of the generative grammar that underlies the observed use of language. Such a study will concern itself with the relation between the data available to the organism and the competence that it acquires; only to the extent that the abstraction to competence has been successful—in the case of language, to the extent that the postulated grammar is "descriptively adequate" in the sense described in Lecture 2 [not reprinted here]—can the investigation of learning hope to achieve meaningful results. If, in some domain, the organization of the behavioral repertoire is quite trivial and elementary, then there will be little harm in avoiding the intermediate stage of theory construction, in which we attempt to characterize accurately the competence that is acquired. But one cannot count on this being the case, and in the study of language it surely is not the case. With a richer and more adequate characterization of "what is learned"—of the underlying competence that constitutes the "final state" of the organism being studied—it may be possible to approach the task of constructing a theory of learning that will be much less restricted in scope than modern behavioral psychology has proved to be. Surely it is pointless to accept methodological strictures that preclude such an approach to problems of learning. . . .

INNATE FACTORS IN LANGUAGE LEARNING

I think that if we contemplate the classical problem of psychology, that of accounting for human knowledge, we cannot avoid being struck by the enormous disparity between knowledge and experience—in the case of language, between the generative grammar that expresses the linguistic competence of the native speaker and the meager and degenerate data on the basis of which he has contructed this grammar for himself. In principle the theory of learning should deal with this problem; but in fact it bypasses the problem, because of the conceptual gap that I mentioned earlier. The problem cannot even be formulated in any sensible way until we develop the concept of competence, alongside the concepts of learning and behavior, and apply this concept in some domain. The fact is that this concept has so far been extensively developed and applied only in the study of human language. It is only in this domain that we have at least the first steps toward an account of competence, namely the fragmentary generative grammars that have been constructed for particular languages. As the study of language progresses, we can expect with some confidence that these grammars will be extended in scope and depth, although it will hardly come as a surprise if the first proposals are found to be mistaken in fundamental ways.

Insofar as we have a tentative first approximation to a generative grammar for some language, we can for the first time formulate in a useful way the problem of origin of knowledge. In other words, we can ask the question, What initial structure must be attributed to the mind that enables it to construct such a grammar from the data of sense? Some of the empirical conditions that must be met by any such assumption about innate structure are moderately clear. Thus, it appears to be a species-specific capacity that is essentially independent of intelligence, and we can make a fairly good estimate of the amount of data that is necessary for the task to be successfully accomplished. We know that the grammars that are in fact constructed vary only slightly among speakers of the same language, despite wide variations not only in intelligence but also in the conditions under which language is acquired. As participants in a certain culture, we are naturally aware of the great differences in ability to use language, in knowledge of vocabulary, and so on that result from differences in native ability and from differences in conditions of acquisition; we naturally pay much less attention to the similarities and to common knowledge, which we take for granted. But if we manage to establish the requisite psychic distance, if we actually compare the generative grammar that must be postulated for different speakers of the same language, we find that the similarities that we take for granted are quite marked and that the divergences are few and marginal. What is more, it seems that dialects that are superficially quite remote, even barely intelligible on first contact, share a vast central core of common rules and processes and differ very slightly in underlying structures, which seem to remain invariant through long historical eras. Furthermore, we discover a substantial system of principles that do not vary among languages that are, as far as we know, entirely unrelated.

The central problems in this domain are empirical ones that are, in principle at least, quite straightforward, difficult as they may be to solve in a satisfactory way. We must postulate an innate structure that is rich enough to account for the disparity between experience and knowledge, one that can account for the construction of the empirically justified generative grammars within the given limitations of time and access to data. At the same time, this postulated innate mental structure must not be so rich and restrictive as to exclude certain known languages. There is, in other words, an upper bound and a lower bound on the degree and exact character of the complexity that can be postulated as innate mental structure. The factual situation is obscure enough to leave room for much difference of opinion over the true nature of this innate mental structure that makes acquisition of language possible. However, there seems to me to be no doubt that this is an empirical issue, one that can be resolved by proceeding along the lines that I have just roughly outlined.

My own estimate of the situation is that the real problem for tomorrow is that of discovering an assumption regarding innate structure that is sufficiently rich, not that of finding one that is simple or elementary enough to be "plausible." There is, as far as I can see, no reasonable notion of "plausibility," no a priori insight into what innate structures are permissible, that can guide the search for a "sufficiently elementary assumption." It would be mere dogmatism to maintain without argument or evidence that the mind is simpler in its innate structure than other biological systems, just as it would be mere dogmatism to insist that the mind's organization must necessarily follow certain set principles, determined in advance of investigation and maintained in defiance of any empirical findings. I think that the study of problems of mind has been very definitely hampered by a kind of apriorism with which these problems are generally approached. In particular, the empiricist assumptions that have dominated the study of acquisition of knowledge for many years seem to me to have been adopted quite without warrant and to have no special status among the many possibilities that one might imagine as to how the mind functions. . . .

INNATE STRUCTURES IN GENERAL

Turning to comparative ethology, it is interesting to note that one of its earliest motivations was the hope that through the "investigation of the a priori, of the innate working hypotheses present in subhuman organisms," it would be possible to shed light on the a priori forms of human thought. This formulation of intent is quoted from an early and little-known paper by Konrad Lorenz. Lorenz goes on to express views very much like those Peirce had expressed a generation earlier. He maintains:

> One familiar with the innate modes of reaction of subhuman organisms can readily hypothesize that the a priori is due to hereditary differentiations of the central nervous system which have become characteristic of the species, producing hereditary dispositions to think in certain forms. . . . Most certainly Hume was wrong when he wanted to derive all that is a priori from that which the senses supply to experience, just as wrong as Wundt or Helmholtz who simply explain it as an abstraction from preceding experience. Adaptation of the a priori to the real world has no more originated from "experience" than adaptation of the fin of the fish to the properties of water. Just as the form of the fin is given a priori, prior to any individual negotiation of the young fish with the water, and just as it is this form that makes possible this negotiation, so it is also the case with our forms of perception and categories in their relationship to our negotiation with the real external world through experience. In the case of animals, we find limitations specific to the forms of experience possible for them. We believe we can demonstrate the closest functional and probably genetic relationship between these animal a priori's and our human a priori. Contrary to Hume, we believe, just as did Kant, that a "pure" science of innate forms of human thought, independent of all experience, is possible.

Peirce, to my knowledge, is original and unique in stressing the problem of studying the rules that limit the class of possible theories. Of course, his concept of abduction, like Lorenz's biological a priori, has a strongly Kantian flavor, and all derive from the rationalist psychology that concerned itself with the forms, the limits, and the principles that provide "the sinews and connections" for human thought, that underlie "that infinite amount of knowledge of which we are not always conscious," of which Leibnitz spoke. It is therefore quite natural that we should link these developments to the revival of philosophical grammar, which grew from the same soil as an attempt, quite fruitful and legitimate, to explore one basic facet of human intelligence.

In recent discussion, models and observations derived from ethology have frequently been cited as providing biological support, or at least analogue, to new approaches to the study of human intelligence. I cite these comments of Lorenz's mainly in order to show that this reference does not distort the outlook of at least some of the founders of this domain of comparative psychology.

One word of caution is necessary in referring to Lorenz, now that he has been discovered by Robert Ardrey and Joseph Alsop and popularized as a prophet of doom. It seems to me that Lorenz's views on human aggression have been extended to near absurdity by some of his expositors. It is no doubt true that there are innate tendencies in the human psychic constitution that lead to aggressiveness under specific social and cultural conditions. But there is little reason to suppose that these tendencies are so dominant as to leave us forever tottering on brink of a Hobbesian war of all against all—as, incidentally, Lorenz at least is fully aware, if I read him rightly. Skepticism is certainly in order when a doctrine of man's "inherent aggressiveness" comes to the surface in a society that glorifies competitiveness, in a civilization that has been distinguished by the brutality of the attack that it has mounted against less fortunate peoples. It is fair to ask to what extent the enthusiasm for this curious view of man's nature is attributable to fact and logic and to what extent it merely reflects the limited extent to which the general cultural level has advanced since the days when Clive and the Portuguese explorers taught the meaning of true savagery to the inferior races that stood in their way.

In any event, I would not want what I am saying to be confused with other, entirely different attempts to revive a theory of human instinct. What seems to me important in ethology is its attempt to explore the innate properties that determine how knowledge is acquired and the character of this knowledge. Returning to this theme, we must consider a further question: How did the human mind come to acquire the innate structure that we are led to attribute to it? Not too surprisingly, Lorenz takes the position that this is simply a matter of natural selection. Peirce offers a rather different speculation, arguing that "nature fecundates the mind of man with ideas which, when these ideas grow up, will resemble their father, Nature." Man is "provided with certain natural beliefs that are true" because "certain uniformities . . . prevail throughout the universe, and the reasoning mind is [it]self a product of this universe. These same laws are thus, by logical necessity, incorporated in his own being." Here, it seems clear that Peirce's argument is entirely without force and that it offers little improvement over the preestablished harmony that it was presumably intended to replace. The fact that the mind is a product of natural laws does not imply that it is equipped to understand these laws or to arrive at them by "abduction." There would be no difficulty in designing a device (say, programing a computer) that is a product of natural law, but that, given data, will arrive at any arbitrary absurd theory to "explain" these data.

In fact, the processes by which the human mind achieved its present stage of complexity and its particular form of innate organization are a total mystery, as much so as the analogous questions about the physical or mental organization of any other complex organism. It is perfectly safe to attribute this development to "natural selection," so long as we realize that there is no substance to this assertion, that it amounts to nothing more than a belief that there is some naturalistic explanation for these phenomena. The problem of accounting for evolutionary development is, in some ways, rather like that of explaining successful abduction. The laws that determine possible successful mutation and the nature of complex organisms are as unknown as the laws that determine the choice of hypotheses. With no knowledge of the laws that determine the organization and structure of complex biological systems, it is just as senseless to ask what the "probability" is for the human mind to have reached its present state as it is to inquire into the "probability" that a particular physical theory will be devised. And, as we have noted, it is idle to speculate about laws of learning until we have some indication of what kind of knowledge is attainable—in the case of language, some indication of the constraints on the set of potential grammars.

In studying the evolution of mind, we cannot guess to what extent there are physically possible alternatives to, say, transformational generative grammar, for an organism meeting certain other physical conditions characteristic of humans. Conceivably, there are none—or very few—in which case talk about evolution of the language capacity is beside the point. The vacuity of such speculation, however, has no bearing one way or another on those aspects of the problem of mind that can be sensibly pursued. It seems to me that these aspects are, for the moment, the problems illustrated in the case of language by the study of the nature, the use, and the acquisition of linguistic competence.

There is one final issue that deserves a word of comment. I have been using mentalistic terminology quite freely, but entirely without prejudice as to the question of what may be the physical realization of the abstract mechanisms postulated to account for the phenomena of behavior or the acquisition of knowledge. We are not constrained, as was Descartes, to postulate a second substance when we deal with phenomena that are not expressible in terms of matter in motion, in his sense. Nor is there much point in pursuing the question of psychophysical parallelism, in this connection. It is an interesting question whether the functioning and evolution of human mentality can be accomodated within the framework of physical explanation, as presently conceived, or whether there are new principles, now unknown, that must be invoked, perhaps principles that emerge only at higher levels of organization than can now be submitted to physical investigation. We can, however, be fairly sure that there will be a physical explanation for the phenomena in question, if they can be explained at all, for an interesting terminological reason, namely that the concept of "physical explanation" will no doubt be extended to incorporate whatever is discovered in this domain, exactly as it was extended to accommodate gravitational and electromagnetic force, massless particles, and numerous other entities and processes that would have offended the common sense of earlier generations. But it seems clear that this issue need not delay the study of the topics that are not open to investigation, and it seems futile to speculate about matters so remote from present understanding.

I have tried to suggest that the study of language may very well, as was traditionally supposed, provide a remarkably favorable perspective for the study of human mental processes. The creative aspect of language use, when investigated with care and respect for the facts, shows that current notions of habit and generalization, as determinants of behavior or knowledge, are quite inadequate. The abstractness of linguistic structure reinforces this conclusion, and it suggests further that in both perception and learning the mind plays an active role in determining the character of the acquired knowledge. The empirical study of linguistic universals has led to the formulation of highly restrictive and, I believe, quite plausible hypotheses concerning the possible variety of human languages, hypotheses that contribute to the attempt to develop a theory of acquisition of knowledge that gives due place to intrinsic mental activity. It seems to me, then, that the study of language should occupy a central place in general psychology.

Surely the classical questions of language and mind receive no final solution, or even the hint of a final solution, from the work that is being actively pursued today. Nevertheless, these problems can be formulated in new ways and seen in a new light. For the first time in many years, it seems to me, there is some real opportunity for substantial progress in the study of the contribution of the mind to perception and the innate basis for acquisition of knowledge. Still, in many respects, we have not made the first approach to a real answer to the classical problems. For example, the central problems relating to the creative aspect of language use remain as inaccessible as they have always been. And the study of universal semantics, surely crucial to the full investigation of language structure, has barely advanced since the medieval period. Many other critical areas might be mentioned where progress has been slow or nonexistent. Real progress has been made in the study of mechanisms of language, the formal principles that make possible the creative aspect of language use and that determine the phonetic form and semantic content of utterances. Our understanding of these mechanisms, though only fragmentary, does seem to me to have real implications for the study of human psychology. By pursuing the kinds of research that now seem feasible and by focusing attention on certain problems that are now accessible to study, we may be able to spell out in some detail the elaborate and abstract computations that determine, in part, the nature of percepts and the character of the knowledge that we can acquire—the highly specific ways of interpreting phenomena that are, in large measure, beyond our consciousness and control and that may be unique to man.

E. O. Wilson

The Harvard biologist Edward O. Wilson has recently claimed to have founded a new discipline, as the title of his book *Sociobiology: The New Synthesis* (1975) indicates. His assertion that human social behavior is significantly determined by a genetic component has been hotly contested by those who follow Marx or Skinner, in favoring explanations in terms of social environment. Obviously, ideological prejudices—in one way or another—have contributed to the passion with which the views of Marx, Freud, Skinner, Lorenz, Chomsky, Wilson, and others have been debated. There are also conceptual ambiguities that cloud the issues—for instance, the notion of genetic determination needs the kind of precise elucidation that Wilson attempts here. But it can be hardly denied that the main question as to which elements in human nature are innate to the species, and which are changeable by different kinds of social influence, is one of fact. There are practical difficulties in determining the facts here, especially because we do not allow ourselves the same liberties in experimenting with human beings as we commonly do with animals. But the evidence for innate factors adduced by Lorenz, Chomsky, Wilson, and many others is impressive. The long search for a scientific theory of human nature has come up with some substantial results.

These excerpts are from Wilson's book *On Human Nature* (Harvard University Press, 1978).

THE GENETIC DETERMINANTS IN HUMAN SOCIAL BEHAVIOR

Thousands of species are highly social. The most advanced among them constitute what I have called the three pinnacles of social evolution in animals: the corals, bryozoans, and other colony-forming invertebrates; the social insects, including ants, wasps, bees, and termites; and the social fish, birds, and mammals. The communal beings of the three pinnacles are among the principal objects of the new discipline of sociobiology, defined as the systematic study of the biological basis of all forms of social behavior, in all kinds of organisms, including man. The enterprise has old roots. Much of its basic information and some of its most vital ideas have come from ethology, the study of whole patterns of behavior of organisms under natural conditions. Ethology was pioneered by Julian Huxley, Karl von Frisch, Konrad Lorenz, Nikolaas Tinbergen, and a few others and is now being pursued by a large new generation of innovative and productive investigators. It has remained most concerned with the particularity of the behavior patterns shown by each species, the ways these patterns adapt animals to the special challenges of their environments, and the steps by which one pattern gives rise to another as the species themselves undergo genetic evolution. Increasingly, modern ethology is being linked to studies of the nervous system and the effects of hormones on behavior. Its investigators have become deeply involved with developmental processes and even learning, formerly the nearly exclusive domain of psychology, and they have begun to include man among the species most closely scrutinized. The emphasis of ethology remains on the individual organism and the physiology of organisms.

Sociobiology, in contrast, is a more explicitly hybrid discipline that incorporates knowledge from ethology (the naturalistic study of whole patterns of behavior), ecology (the study of the relationships of organisms to their environment), and genetics in order to derive general principles concerning the biological properties of entire societies. What is truly new about sociobiology is the way it has extracted the most important facts about social organization from their traditional matrix of ethology and psychology and reassembled them on a foundation of ecology and genetics studied at the population level in order to show how social groups adapt to the environment by evolution. Only within the past few years have ecology and genetics themselves become sophisticated and strong enough to provide such a foundation.

Sociobiology is a subject based largely on comparisons of social species. Each living form can be viewed as an evolutionary experiment, a product of millions of years of interaction between genes and environment. By examining many such experiments closely, we have begun to construct and test the first general principles of genetic social evolution. It is now within our reach to apply this broad knowledge to the study of human beings.

Sociobiologists consider man as though seen through the front end of a telescope, at a greater than usual distance and temporarily diminished in size, in order to view him simultaneously with an array of other social experiments. They attempt to place human-kind in its proper place in a catalog of the social species on Earth. They agree with Rousseau that "One needs to look near at hand in order to study men, but to study man one must look from afar." . . .

Human behavior is something that can be defined with fair precision, because the evolutionary pathways open to it have not all been equally negotiable. Evolution has not made culture all-powerful. It is a misconception among many of the more traditional Marxists, some learning theorists, and a still surprising proportion of anthropologists and sociologists that social behavior can be shaped into virtually any form. Ultra-environmentalists start with the premise that man is the creation of his own culture: "culture makes man," the formula might go, "makes culture makes man." Theirs is only a half truth. Each person is molded by an interaction of his environment, especially his culture environment, with the genes that affect social behavior. Although the hundreds of the world's cultures seem enormously variable to those of us who stand in their midst, all versions of human social behavior together form only a tiny fraction of the realized organizations of social species on this planet and a still smaller fraction of those that can be readily imagined with the aid of sociobiological theory.

The question of interest is no longer whether human social behavior is genetically determined; it is to what extent. The accumulated evidence for a large hereditary component is more detailed and compelling than most persons, including even geneticists, realize. I will go further; it already is decisive.

That being said, let me provide an exact definition of a genetically determined trait. It is a trait that differs from other traits at least in part as a result of the presence of one or more distinctive genes. The important point is that the objective estimate of genetic influence requires comparison of two or more states of the same feature. To say that blue eyes are inherited is not meaningful without further qualification, because blue eyes are the product of an interaction between genes and the largely physiological environment that brought final coloration to the irises. But to say that the *difference* between blue and brown eyes is based wholly or partly on differences in genes is a meaningful statement because it can be tested and translated into the laws of genetics. Additional information is then sought: What are the eye colors of the parents, siblings, children, and more distant relatives? These data are compared to the very simplest model of Mendelian heredity, which, based on our understanding of cell multiplication and sexual reproduction, entails the action of only two genes. If the data fit, the differences are interpreted as being based on two genes. If not, increasingly complicated schemes are applied. Progressively larger numbers of genes and more complicated modes of interaction are assumed until a reasonably close fit can be made. In the example just cited, the main differences between blue and brown eyes are in fact based on two genes, although complicated modifications exist that make them less than an ideal textbook example. In the case of the most complex traits, hundreds of genes are sometimes involved, and their degree of influence can ordinarily be measured only crudely and with the aid of sophisticated mathematical techniques. Nevertheless, when the analysis is properly performed it leaves little doubt as to the presence and approximate magnitude of the genetic influence.

Human social behavior can be evaluated in essentially the same way, first by comparison with the behavior of other species and then, with far greater difficulty and ambiguity, by studies of variation among and within human populations. The picture of genetic determinism emerges most sharply when we compared selected major categories of animals with the human species. Certain general human traits are shared with a majority of the great apes and monkeys of Africa and Asia, which on grounds of anatomy and biochemistry are our closest living evolutionary relatives:

• Our intimate social groupings contain on the order of ten to one hundred adults, never just two, as in most birds and marmosets, or up to thousands, as in many kinds of fishes and insects.

• Males are larger than females. This is a characteristic of considered significance within the Old World monkeys and apes and many other kinds of mammals. The average number of females consorting with successful males closely corresponds to the size gap between males and females when many species are considered together. The rule makes sense: the greater the competition among males for females, the greater the advantage of large size and the less influential are any disadvantages accruing to bigness. Men are not very much larger than women; we are similar to chimpanzees in this regard. When the sexual size difference in human beings is plotted on the curve based on other kinds of mammals, the predicted average number of females per successful male turns out to be greater than one but less than three. The prediction is close to reality, we know we are a mildly polygynous species.

• The young are molded by a long period of social training, first by closest associations with the mother, then to an increasing degree with other children of the same age and sex.

• Social play is a strongly developed activity featuring role practice, mock aggression, sex practice, and exploration.

These and other properties together identify the taxonomic group consisting of Old World monkeys, the great apes, and human beings. It is inconceivable that human beings could be socialized into the radically different repertories of other groups such as fishes, birds, antelopes, or rodents. Human beings might self-consciously *imitate* such arrangements, but it would be a fiction played out on a stage, would run counter to deep emotional responses and have no chance of persisting through as much as a single generation. To adopt with serious intent, even in broad outline, the social system of a nonprimate species would be insanity in the literal sense. Personalities would quickly dissolve, relationships disintegrate, and reproduction cease.

At the next, finer level of classification, our species is distinct from the Old World monkeys and apes in ways that can be explained only as a result of a unique set of human genes. Of course, that is a point quickly conceded by even the most ardent environmentalists. They are willing to agree with the great geneticist Theodosius Dobzhansky that "in a sense, human genes have surrendered their primacy in human evolution to an entirely new, nonbiological or superorganic agent, culture. However, it should not be forgotten that this agent is entirely dependent on the human genotype." But the matter is much deeper and more interesting than that. There are

social traits occurring through all cultures which upon close examination are as diagnostic of mankind as are distinguishing characteristics of other animal species—as true to the human type, say, as
wing tessellation is to a fritillary butterfly or a complicated spring
melody to a wood thrush. In 1945 the American anthropologist
George P. Murdock listed in the following characteristics that have
been recorded in every culture known to history and ethnography:

> Age-grading, athletic sports, bodily adornment, calendar, cleanliness
> training, community organization, cooking, cooperative labor, cos-
> mology, courtship, dancing, decorative art, divination, division of la-
> bor, dream interpretation, education, eschatology, ethics, ethnobot-
> any, etiquette, faith healing, family feasting, fire making, folklore,
> food taboos, funeral rites, games, gestures, gift giving, government,
> greetings, hair styles, hospitality, housing, hygiene, incest taboos, in-
> heritance rules, joking, kin groups, kinship nomenclature, language,
> law, luck superstitions, magic, marriage, mealtimes, medicine, obstet-
> rics, penal sanctions, personal names, population policy, postnatal
> care, pregnancy usages, property rights, propitiation of supernatural
> beings, puberty customs, religious ritual, residence rules, sexual re-
> strictions, soul concepts, status differentiation, surgery, tool making,
> trade, visiting, weaving, and weather control.

Few of these unifying properties can be interpreted as the inevitable outcome of either advanced social life or high intelligence. It is
easy to imagine nonhuman societies whose members are even more
intelligent and complexly organized than ourselves, yet lack a majority of the qualities just listed. . . .

To summarize the argument to this point: the general traits of
human nature appear limited and idiosyncratic when placed against
the great backdrop of all other living species. Additional evidence
suggests that the more stereotyped forms of human behavior are
mammalian and even more specifically primate in character, as predicted on the basis of general evolutionary theory. Chimpanzees are
close enough to ourselves in the details of their social life and mental properties to rank as nearly human in certain domains where it
was once considered inappropriate to make comparisons at all.
These facts are in accord with the hypothesis that human social
behavior rests on a genetic foundation—that human behavior is, to

be more precise, organized by some genes that are shared with closely related species and others that are unique to the human species. The same facts are unfavorable for the competing hypothesis which has dominated the social sciences for generations, that mankind has escaped its own genes to the extent of being entirely culture-bound.

Let us pursue this matter systematically. The heart of the genetic hypothesis is the proposition derived in a straight line from neo-Darwinian evolutionary theory, that the traits of human nature were adaptive during the time that the human species evolved and that genes consequently spread through the population that predisposed their carriers to develop those traits. Adaptiveness means simply that if an individual displayed the traits he stood a greater chance of having his genes represented in the next generation than if he did not display the traits. The differential advantage among individuals in this strictest sense is called genetic fitness. There are three basic components of genetic fitness: increased personal survival, increased personal reproduction, enhanced survival and reproduction of close relatives who share the same genes by common descent. An improvement in any one of the factors or in any combination of them results in greater genetic fitness. The process, which Darwin called natural selection, describes a tight circle of causation. If the possession of certain genes predisposes individuals toward a particular trait, say a certain kind of social response, and the trait in turn conveys a superior fitness, the genes will gain an increased representation in the next generation. If natural selection is continued over many generations, the favored genes will spread throughout the population, and the trait will become characteristic of the species. In this way human nature is postulated by many sociobiologists, anthropologists, and others to have been shaped by natural selection. . . .

We can be fairly certain that most of the genetic evolution of human social behavior occurred over the five million years prior to civilization, when the species consisted of sparse, relatively immobile populations of hunter-gathers. On the other hand, by far the greater part of cultural evolution has occurred since the origin of agriculture and cities approximately 10,000 years ago. Although genetic evolution of some kind continued during this latter, historical sprint, it cannot have fashioned more than a tiny fraction of the traits of human nature. Otherwise surviving hunter-gatherer people would differ genetically to a significant degree from people in advanced industrial nations, but this is demonstrably not the case. It follows that human sociobiology can be most directly tested in studies of hunter-gatherer societies and the most persistent preliterate herding and agriculture societies. As a result, anthropology rather than sociology or economics is the social science closest to sociobiology. It is in anthropology that the genetic theory of human nature can be most directly pursued.

The power of a scientific theory is measured by its ability to transform a small number of axiomatic ideas into detailed predictions of observable phenomena; thus the Bohr atom made modern chemistry possible, and modern chemistry recreated cell biology. Further, the validity of a theory is measured by the extent to which its predictions successfully compete with other theories in accounting for the phenomena; the solar system of Copernicus won over that of Ptolemy, after a brief struggle. Finally, a theory waxes in influence and esteem among scientists as it assembles an ever larger body of facts into readily remembered and usable explanatory schemes, and as newly discovered facts conform to its demands: the round earth is more plausible than a flat one. Facts crucial to the advancement of science can be obtained either by experiments designed for the purpose of acquiring them or from the inspired observation of undisturbed natural phenomena. Science has always progressed in approximately this opportunistic, zig-zagging manner.

In the case of the theory of the genetic evolution of human nature, if it is ever to be made part of real science, we should be able to select some of the best principles from ecology and genetics, which are themselves based on the theory, and adapt them in detail to human social organization. The theory must not only account for many of the known facts in a more convincing manner than traditional explanations, but must also identify the need for new kinds of information previously unimagined by the social sciences. The behavior thus explained should be the most general and least rational of the human repertoire, the part furthest removed from the influence of day-to-day reflection and the distracting vicissitudes of culture. In other words, they should implicate innate, biological phenomena that are the least susceptible to mimicry by culture.

These are stern requirements to impose on the infant discipline of human sociobiology, but they can be adequately justified. Sociobiology intrudes into the social with credentials from the natural sciences and, initially, an unfair psychological advantage. If the ideas and analytical methods of "hard" science can be made to work in a congenial and enduring manner, the division between the two cultures of science and the humanities will close. But if our conception of human nature is to be altered, it must be by means of truths conforming to the canons of scientific evidence and not a new dogma however devoutly wished for.

Part IV

How Far Is a Scientific Theory of Human Nature Possible? Twentieth-Century Philosophical Views

We have been following a trail that has pursued the idea of a scientific account of human nature from the seventeenth century right up to the present day. The trail started with the philosophers trying to prove conclusions by reason alone (of whom Spinoza was the purest example), but it has led us to thinkers such as Lorenz and Wilson, who would certainly wish to describe themselves as scientists, aiming to test their theories against observable fact. Along the way, however, we have noted various doubts about whether the methods of the physical sciences can be applied to *all* questions about human nature.

Now it is time to take up, in a little more detail, these philosophical problems that have been with us all along; to ask about the forms that sciences of human nature can take, and the limits to which they are subject. That there are facts about human nature to be discovered by the methods of the established physical sciences is beyond doubt, for the disciplines of physiology, biochemistry, genetics, and so forth, are steadily adding to our knowledge. The controversy is over whether the so-called "social sciences" such as psychology, linguistics, economics, sociology, among others, can or should follow just the same methodology as the physical sciences (as recommended by Mill).

It may be said that the social sciences all aim to explain human behavior, but this form of words does not settle the issue, because it remains controversial what the relevant concepts of "behavior" and of "explanation" should be. In most of the social sciences, what seems essential is not merely the physically measurable movements of human bodies, but the intentional action thereby performed (for example, the purchase made, not just the handing over of coin; the perceptual judgment asserted, not just the noises made). But the ascriptions of intentions in acting involves, as we shall see, the ascription of a large range of other mental states, including beliefs and desires. Thus questions quickly arise of what *justifies* us in ascribing such mental states to people, and of what such states really *are*—in particular, of how they relate to physically describable states of the body, especially the brain. The concept of explanation also raises deep philosophical problems—whether giving causes of an action thereby rules out the action as genuinely free or voluntary or even intentional, and whether in our everyday explanation of actions in terms of the agent's reasons we are giving *any* sort of causal explanation. No doubt, it is because of the difficulty of these philosophical problems that the status of psychoanalytic theorizing remains unclear and controversial.

These problems have been given considerable attention in the various schools of twentieth-century philosophy—represented here by Sartre (existentialist and phenomenological), Hampshire, Feigl, and Davidson (English-speaking analytical philosophy— though Feigl was formerly a member of the Vienna circle of logical positivists), and Habermas (the German traditions of the *Geisteswissenschaften* and neo-Marxist critical theory). Of course, these brief extracts can be no more than a shop window for a vast range of philosophical goods.

Jean-Paul Sartre

Jean-Paul Sartre is famous as a philosopher, novelist, playwright, and biographer, and as the figure at the center of French intellectual life in the postwar generation. His name will forever be associated with existentialism—that movement of thought which places supreme value on the freedom of the individual and the meaning of life. The movement includes such diverse figures as the eccentric Danish Protestant Kierkegaard, the atheist Nietszche, who joyfully proclaimed the death of God, and Heidegger, who explored what are supposed to be the concrete realities of the human condition in the ponderous abstractions of *Being and Time*. Sartre's philosophical style, as will be evident even in these excerpts, suffers from the influence of Heidegger, but his thought repays the effort of understanding it demands.

In the context of this anthology, Sartre's position stands out as a profound analysis, and resolute assertion, of man's essential freedom as something not open to any kind of scientific explanation. Sartre shows no interest in the possible survival of a disembodied soul, but his view of human nature involves a very Cartesian, sharp distinction between being-for-itself (consciousness) and being-in-itself (material things). In his analysis, consciousness and freedom are inseparable. His discussion of action provides a convenient starting point for these extracts. He emphasizes that all intentional action must by definition be an attempt to change the world in some way, to bring about what the agent believes to be desirable, but *not* already the case. So it implies the ability to make such negative judgments, which Sartre insists on treating in the mystifying terms of "Nothingness," but which we can understand in terms of what Brentano called *intentionality*—the ability to believe, consider, desire, fear, etc., what may or may not actually be the case. Sartre's claim seems to be that the reality of human freedom is proved by our having this ability.

He goes on to argue that there is a mode of consciousness, which he calls by the typically dramatic name of "anguish," in which we are peculiarly aware of our freedom, of the insufficiency of all desires, decisions, or values to determine our conduct. But we normally do our best to avoid this unsettling awareness and live most of our lives in what Sartre calls "bad faith," pretending (even to ourselves) not to be free. He rejects

Freudian accounts of bad faith as themselves too mechanistic, avoiding the problem only by transferring it to a postulated mental element (the censor) in the unconscious. His own treatment, however, seems to leave us with the self-contradiction that human reality must "be what it is not and not be what it is." The next reading from Stuart Hampshire tries to take the discussion further.

These excerpts are from *Being and Nothingness* (first published in 1943), Part One, Chapters 1 and 2, Part Four, Chapter 1. The translation is by Hazel Barnes (Methuen, 1957; in the United States, Washington Square Press, a Division of Simon & Schuster).

INTENTION AND ACTION

It is strange that philosophers have been able to argue endlessly about determinism and free-will, to cite examples in favor of one or the other thesis without ever attempting first to make explicit the structures contained in the very idea of *action*. The concept of an act contains, in fact, numerous subordinate notions which we shall have to organize and arrange in a hierarchy: to act is to modify the shape of the world; it is to arrange means in view of an end; it is to produce an organized instrumental complex such that by a series of concatenations and connections the modification effected on one of the links causes modifications throughout the whole series and finally produces an anticipated result. But this is not what is important for us here. We should observe first that an action is on principle *intentional*. The careless smoker who has through negligence caused the explosion of a powder magazine has not *acted*. On the other hand the worker who is charged with dynamiting a quarry and who obeys the given orders has acted when he has produced the expected explosion; he knew what he was doing or, if you prefer, he intentionally realized a conscious project.

This does not mean, of course, that one must foresee all the consequences of his act. The emperor Constantine when he established himself at Byzantium, did not foresee that he would create a center of Greek culture and language, the appearance of which would ultimately provoke a schism in the Christian Church and which would contribute to weakening the Roman Empire. Yet he

performed an act just in so far as he realized his project of creating a
new residence for emperors in the Orient. Equating the result with
the intention is here sufficient for us to be able to speak of action.
But if this is the case, we establish that the action necessarily implies
as its condition the recognition of a "desideratum"; that is, of an
objective lack or again of a *négatité*. The *intention* of providing a
rival for Rome can come to Constantine only through the apprehension of an objective lack: Rome lacks a counterweight; to this
still profoundly pagan city ought to be opposed a Christian city
which at the moment *is missing*. Creating Constantinople is understood as an act only if first the conception of a new city has preceded the action itself or at least if this conception serves as an
organizing theme for all later steps. But this conception can not be
the pure representation of the city as *possible*. It apprehends the city
in its essential characteristic, which is to be a *desirable* and not yet
realized possible.

This means that from the moment of the first conception of the
act, consciousness has been able to withdraw itself from the full
world of which it is consciousness and to leave the level of being in
order frankly to approach that of non-being. Consciousness in so
far as it is considered exclusively in its being, is perpetually referred
from being to being and can not find in being any motive for
revealing non-being. The imperial system with Rome as its capital
functions positively and in a certain real way which can be easily
discovered. Will someone say that the taxes are collected badly, that
Rome is not secure from invasions, that it does not have the geographical location which is suitable for the capital of a Mediterra-
nean empire which is threatened by barbarians, that its corrupt
morals make the spread of the Christian religion difficult? How can
anyone fail to see that all these considerations are *negative;* that is,
that they aim at what is not, not at what is. To say that sixty per
cent of the anticipated taxes have been collected can pass, if need be
for a positive appreciation of the situation such as it is. To say that
they are *badly* collected is to consider the situation across a situation
which is posited as an absolute end but which precisely is not. To
say that the corrupt morals at Rome hinder the spread of Christianity is not to consider this diffusion for what it is; that is, for a
propagation at a rate which the reports of the clergy can enable us

to determine. It is to posit the diffusion in itself as insufficient; that is, as suffering from a secret of nothingness. But it appears as such only if it is surpassed toward a limiting-situation posited *a priori* as a value (for example, toward a certain rate of religious conversions, toward a certain mass morality). This limiting-situation can not be conceived in terms of the simple consideration of the real state of things; for the most beautiful girl in the world can offer only what she *has,* and in the same way the most miserable situation can by itself be designated only as it is without any reference to an ideal nothingness.

In so far as man is immersed in the historical situation, he does not even succeed in conceiving of the failures and lacks in a political organization or determined economy; this is not, as is stupidly said, because he 'is accustomed to it," but because he apprehends it in its plenitude of being and because he can not even imagine that he can exist in it otherwise. For it is necessary here to reverse common opinion and on the basis of what it is not, to acknowledge the harshness of a situation or the sufferings which it imposes, both of which are motives for conceiving of another state of affairs in which things would be better for everybody. It is on the day that we can conceive of different state of affairs that a new light falls on our troubles and our suffering and that we *decide* that these are unbearable. A worker in 1830 is capable of revolting if his salary is lowered, for he easily conceives of a situation in which his wretched standard of living would be not as low as the one which is about to be imposed on him. But he does not represent his sufferings to himself as unbearable; he adapts himself to them not through resignation but because he lacks the education and reflection necessary for him to conceive of a social state in which these sufferings would not exist. Consequently *he does not act.* Masters of Lyon following a riot, the workers at Croix-Rousse do not know what to do with their victory; they return home bewildered, and the regular army has no trouble in overcoming them. Their misfortunes do not appear to them "habitual" but rather *natural;* they *are,* that is all, and they constitute the worker's condition. They are not detached; they are not seen in the clear light of day, and consequently, they are integrated by the worker with his being. He suffers without considering his suffering and without conferring value upon it. To suffer and to be are one and the same for him. His suffering is the pure affective tenor of his nonpositional consciousness, but he does not *contemplate* it. Therefore this suffering can not be in itself a *motive* for his acts. Quite the contrary, it is after he has formed the project of changing the situation that it will appear intolerable to him. This means that he will have had to give himself room, to withdraw in relation to it, and will have to have effected a double nihilation: on the one hand, he must posit an ideal state of affairs as a pure *present* nothingness; on the other hand, he must posit the actual situation as nothingness in relation to this state of affairs. He will have to conceive of a happiness attached to his class as a pure possible—that is, presently as a certain nothingness—and on the other hand, he will return to the present situation in order to illuminate it in the light of this nothingness and in order to nihilate it in turn by declaring: "I *am not* happy."

Two important consequences result. (1) No factual state whatever it may be (the political and economic structure of society, the psychological "state," etc.) is capable by itself of motivating any act whatsoever. For an act is a projection of the for-itself toward what is not, and what is can in no way determine by itself what is not. (2) No factual state can determine consciousness to apprehend it as a *négatité* or as a lack. Better yet no factual state can determine consciousness to define it and circumscribe it since, as we have seen, Spinoza's statement, "Omnis determinatio est negatio," remains profoundly true. Now every action has for its express condition not only the discovery of a state of affairs as "lacking in——," that is, as a *négatité*—but also, and before all else, the constitution of the state of things under consideration into an isolated system. There is a factual state—satisfying or not—only by means of the nihilating power of the for-itself. But this power of nihilation can not be limited to realizing a simple *withdrawal* in relation to the world. In fact in so far as consciousness is "invested" by being, in so far as it simply suffers what is, it must be included in being. It is the organized form—worker-finding-his-suffering-natural—which must be surmounted and denied in order for it to be able to form the object of a revealing contemplation. This means evidently that it is by a pure wrenching away from himself and the world that the worker can posit his suffering as unbearable suffering and consequently can *make of it of the motive* for his revolutionary action. This implies for consciousness the permanent possibility of effecting a rupture with its own past, of wrenching itself away from its past so as to be able to consider it in the light of a non-being and so as to be able to confer on it the meaning which *it has* in terms of the project of a meaning which it *does not have*. Under no circumstances can the past in any way by itself produce *an act;* that is, the positing of an end which turns back upon itself so as to illuminate it. This is what Hegel caught sight of when he wrote that "the mind is the negative," although he seems not to have remembered this when he came to presenting his own theory of action and of freedom. In fact as soon as one attributes to consciousness this negative power with respect to the world and itself, as soon as the nihilation forms an integral part of the *positing* of an end, we must recognize that the indispensable and fundamental condition of all action is the freedom of the acting being. . . .

NOTHINGNESS AND FREEDOM

. . . Every question in essence posits the possibility of a negative reply. In a question we question a being about its being or its way of being. This way of being or this being is veiled; there always remains the possibility that it may unveil itself as a Nothingness. But from the very fact that we presume that an Existent can always be revealed as *nothing,* every question supposes that we realize a nihilating withdrawal in relation to the given, which becomes a simple *presentation,* fluctuating between being and Nothingness.

It is essential therefore that the questioner have the permanent possibility of dissociating himself from the causal series which constitutes being and which can produce only being. If we admitted that the question is determined in the questioner by universal determinism, the question would thereby become unintelligible and even inconceivable. A real cause, in fact, produces a real effect and the caused being is wholly engaged by the cause in positivity; to the extent that its being depends on the cause, it can not have within itself the tiniest germ of nothingness. Thus in so far as the questioner must be able to effect in relation to the questioned a kind of nihilating withdrawal, he is not subject to the causal order of the world; he detaches himself from Being. This means that by a double movement of nihilation, he nihilates the thing questioned in relation to himself by placing it in a *neutral* state, between being and non-being—and that he nihilates himself in relation to the thing questioned by wrenching himself from being in order to be able to bring out of himself the possibility of a non-being. Thus in posing a question, a certain negative element is introduced into the world. We see nothingness making the world irridescent, casting a shimmer over things. But at the same time the question emanates from a questioner who in order to motivate himself in his being as one who questions, disengages himself from being. This disengagement is then by definition a human-process. Man presents himself at least in this instance as a being who causes Nothingness to arise in the world, inasmuch as he himself is affected with non-being to this end.

These remarks may serve as guiding thread as we examine the *négatités* of which we spoke earlier. There is no doubt at all that these are transcendent realities; distance, for example, is imposed on us as something which we have to take into account, which must be cleared with effort. However these realities are of a very peculiar nature; they all indicate immediately an essential relation of human reality to the world. They derive their origin from an act, an expectation, or a project of the human being; they all indicate an aspect of of being as it appears to the human being who is engaged in the world. The relations of man in the world, which the *négatités* indicate, have nothing in common with the relations *a posteriori* which are brought out by empirical activity. We are no longer dealing with those relations of *instrumentality* by which, according to Heidegger, objects in the world disclose themselves to "human reality." Every *négatité* appears rather as one of the essential conditions of this relation of instrumentality. In order for the totality of being to order itself around us as instruments, in order for it to parcel itself into differentiated complexes which refer one to another and which can *be used*, it is necessary that negation rise up not as a thing among other things but as the rubric of a category which presides over the arrangement and the redistribution of great masses of being in things. Thus the rise of man in the midst of the being which "invests" him causes a world to be discovered. But the essential and primordial moment of this rise is the negation. Thus we have reached the first goal of this study. Man is the being through whom nothingness comes to the world. But this question immediately provokes another: What must man be in his being in order that through him nothingness may come to being?

Being can generate only being and if man is inclosed in this process of generation, only being will come out of him. If we are to assume that man is able to question this process—that is, to make it the object of interrogation—he must be able to hold it up to view as a totality. He must be able to put himself outside of being and by the same stroke weaken the structure of the being of being. Yet it is not given to "human reality" to annihilate even provisionally the mass of being which it posits before itself. Man's *relation* with being is that he can modify it. For man to put a particular existent out of circuit is to put himself out of circuit in relation to that existent. In this case he is not subject to it; he is out of reach; it can not act on him, for he has retired *beyond a nothingness*. Descartes following the Stoics has given a name to this possibility which human reality has to secrete a nothingness which isolates it—it is *freedom*. But freedom here is only a name. If we wish to penetrate further into the question, we must not be content with this reply and we ought to ask now, What is human freedom if through it nothingness comes into the world?

It is not yet possible to deal with the problem of freedom in all its fullness. In fact the steps which we have completed up to now show clearly that freedom is not a faculty of the human soul to be envisaged and described in isolation. What we have been trying to define is the being of man in so far as he conditions the appearance of nothingness, and this being has appeared to us as freedom. Thus freedom as the requisite condition for the nihilation of nothingness is not a *property* which belongs among others to the essence of the human being. We have already noticed furthermore that with man the relation of existence to essence is not comparable to what it is for the things of the world. Human freedom precedes essence in man and makes it possible; the essence of human being is suspended in his freedom. What we call freedom is impossible to distinguish from the *being* of "human reality." Man does not exist *first* in order to be free *subsequently;* there is no difference between the being of man and his *being-free.* . . .

ANGUISH AS CONSCIOUSNESS OF FREEDOM

If freedom is the being of consciousness, consciousness ought to exist as a consciousness of freedom. What form does this consciousness of freedom assume? In freedom the human being is his own past (as also his own future) in the form of nihilation. If our analysis has not led us astray, there ought to exist for the human being, in so far as he is conscious of being, a certain mode of standing opposite his past and his future, as being both this past and this future and as not being them. We shall be able to furnish an immediate reply to this question; it is in anguish that man gets the consciousness of his freedom, or if you prefer, anguish is the mode of being of freedom as consciousness of being; it is in anguish that freedom is, in its being, in question for itself.

Kierkegaard describing anguish in the face of what one lacks characterizes it as anguish in the face of freedom. But Heidegger, whom we know to have been greatly influenced by Kierkegaard, considers anguish instead as the apprehension of nothingness. These two descriptions of anguish do not appear to us contradictory; on the contrary the one implies the other.

First we must acknowledge the Kierkegaard is right; anguish is distinguished from fear in that fear is fear of beings in the world whereas anguish is anguish before myself. Vertigo is anguish to the extent that I am afraid not of falling over the precipice, but of throwing myself over. A situation provokes fear if there is a possibility of my life being changed from without; my being provokes anguish to the extent that I distrust myself and my own reactions in that situation. The artillery preparation which precedes the attack can provoke fear in the soldier who undergoes the bombardment, but anguish is born in him when he tries to foresee the conduct with which he will face the bombardment, when he asks himself if he is going to be able to "hold up." Similarly the recruit who reports for active duty at the beginning of the war can in some instances be afraid of death but more often he is "afraid of being afraid"; that is, he is filled with anguish before himself. Most of the time dangerous or threatening situations present themselves in facets; they will be apprehended through a feeling of fear or anguish according to whether we envisage the situation as acting on the man or the man as acting on the situation. The man who has just received a hard blow—for example, losing a great part of his wealth in a crash—can have the fear of threatening poverty. He will experience anguish a moment later when nervously wringing his hands (a symbolic reaction to the action which is imposed but which remains still wholly undetermined), he exclaims to himself; "What am I going to do? But what am I going to do?" In this sense fear and anguish are exclusive of one another since fear is unreflective apprehension of the transcendent and anguish is reflective apprehension of the self; the one is born in the destruction of the other. The normal process in the case which I have just cited is a constant transition from the one to the other. But there exist also situations where anguish appears pure; that is, without ever being preceded or followed by fear. If, for example, I have been raised to a new dignity and charged with a delicate and flattering mission, I can feel anguish at the thought that I will not be capable perhaps of fulfilling it, and yet I will not have the least fear in the world of the consequences of my possible failure.

What is the meaning of anguish in the various examples which I have just given? Let us take up again the example of vertigo. Vertigo announces itself through fear: I am on a narrow path—without a guard-rail—which goes along a precipice. The precipice presents itself to me as *to be avoided;* it represents a danger of death. At the same time I conceive of a certain number of causes, originating in universal determinism, which can transform that threat of death into reality; I can slip on a stone and fall into the abyss; the crumbling earth of the path can give way under my steps. Through these various anticipations, I am given to myself as a thing; I am passive in relation to these possibilities; they come to me from without; in so far as I am also an object in the world, subject to gravitation, they are my possibilities. At this moment *fear* appears, which in terms of the situation is the apprehension of myself as a destructible transcendent in the midst of transcendents, as an object which does not contain in itself the origin of its future disappearance. My reaction will be of the reflective order: I will pay attention to the stones in the road; I will keep myself as far as possible from the edge of the path. I realize myself as pushing away the threatening situation with all my strength, and I project before myself a certain number of future conducts destined to keep the threats of the world at a distance from me. These conducts are *my* possibilities. I escape fear by the very fact that I am placing myself on a plane where *my own* possibilities are substituted for the transcendent probabilities where human action had no place.

But these conducts, precisely because they are my possibilities, do not appear to me as determined by foreign causes. Not only is it not strictly certain that they will be effective; in particular it is not strictly certain that they will be adopted, for they do not have existence sufficient in itself. We could say, varying the expression of Berkeley, that their "being is a sustained-being" and that their "possibility of being is only an ought-to-be-sustained." Due to this fact their possibility has as a necessary condition the possibility of negative conduct (*not* to pay attention to the stones in the road, to run, to think of something else) and the possibility of the opposite conduct (to throw myself over the precipice). The possibility which I make *my* concrete possibility can appear as my possibility only by raising itself on the basis of the totality of the logical possibilities which the situation allows. But these rejected possibles in turn have no other being then their "sustained-being;" it is I who sustain them in being, and inversely, their present non-being is an "ought-not-to-be-sustained." No external cause will remove them. I alone am the permanent source of their non-being, I engage myself in them; in order to cause *my* possibility to appear, I posit the other possibilities so as to nihilate them. This would not produce anguish if I could apprehend myself in my relations with these possibles as a cause of producing its effects. In this case the effect defined as my possibility *would be strictly* determined. But then it would cease to be *possible;* it would become simply "about-to-happen." If then I wished to avoid anguish and vertigo, it would be enough if I were to consider the motives (instinct of self-preservation, prior fear, etc.), which make me reject the situation envisaged, as *determining* my prior activity in the same way that the presence at a determined point of one given mass determines the courses followed by other masses; it would be necessary, in other words, that I apprehend in myself a strict psychological determinism. But I am in anguish precisely because any conduct on my part is only *possible,* and this means that while constituting a totality of motives *for* pushing away that situation, I at the same moment apprehend these motives as not sufficiently effective. At the very moment when I apprehend my being as *horror* of the precipice, I am conscious of that horror as *not determinant* in relation to my possible conduct. In one sense that horror calls for prudent conduct, and it is in itself a pre-outline of that conduct; in another sense, it posits the final developments of that conduct only as possible, precisely because I do not apprehend it as the cause of these final developments but as need, appeal, etc.

Now as we have seen, consciousness of being is the being of consciousness. There is no question here of a contemplation which I could make after the event, of an horror already constituted; it is the very being of horror to appear to itself as "not being the cause" of the conduct it calls for. In short, to avoid fear, which reveals to me a transcendent future strictly determined, I take refuge in reflection, but the latter has only an undetermined future to offer. This means that in establishing a certain conduct as a possibility and precisely because it is *my* possibility, I am aware that *nothing* can compel me to adopt that conduct. Yet I am indeed already there in the future; it is for the sake of that being which I will be there at the turning of the path that I now exert all my strength, and in this sense there is already a relation between my future being and my present being. But a nothingness has slipped into the heart of this relation; I *am* not the self which I will be. First I am not that self because time separates me from it. Secondly, I am not that self because what I am is not the foundation of what I will be. Finally I am not that self because no actual existent can determine strictly what I am going to be. Yet as I am already what I will be (otherwise I would not be interested in any one being more than another), *I am the self which I will be, in the mode of not being it.* It is through my horror that I am carried toward the future, and the horror nihilates itself in that it constitutes the future as possible. Anguish is precisely my consciousness of being my own future, in the mode of not-being. To be exact, the nihilation of horror as a *motive,* which has the effect of reinforcing horror as a *state,* has as its positive counterpart the appearance of other forms of conduct (in particular that which consists in throwing myself over the precipice) as *my* possible *possibilities.* If *nothing* compels me to save my life, *nothing* prevents me from precipitating myself into the abyss. The decisive conduct will emanate from a self which I am not yet. Thus the self which I am depends on the self which I am not yet to the exact extent that the self which I am not yet does not depend on the self which I am. Vertigo appears as the apprehension of this dependence. I approach the precipice, and my scrutiny is searching for myself in my very depths. In terms of this moment, I play with my possibilities. My eyes, running over the abyss from top to bottom, imitate the possible fall and realize it symbolically; at the same time suicide, from the fact that it becomes a *possibility* possible for *me,* now causes to appear possible motives for adopting it (suicide would cause anguish to cease). Fortunately these motives in their turn, from the sole fact that they are motives of a possibility, present themselves as ineffective, as non-determinant; they can no more *produce* the suicide than my horror of the fall can *determine me* to avoid it. It is this counter-anguish which generally puts an end to anguish by transmuting it into indecision. Indecision in its turn, calls for decision. I abruptly put myself at a distance from the edge of the precipice and resume my way.

The example which we have just analyzed has shown us what we could call "anguish in the face of the future." There exists another: anguish in the face of the past. It is that of the gambler who has freely and sincerely decided not to gamble any more and who when he approaches the gaming table, suddenly sees all his resolutions melt away. This phenomenon has often been described as if the sight of the gaming table reawakened in us a tendency which entered into conflict with our former resolution and ended by drawing us in spite of this. Aside from the fact that such a description is done in materialistic terms and peoples the mind with opposing forces (there is, for example, the moralists' famous "struggle of reason with the passions"), it does not account for the facts. In reality—the letters of Dostoevsky bear witness—there is nothing in us which resembles an inner debate as if we had to

279

weigh motives and incentives before deciding. The earlier resolution of "not playing anymore" is always there, and in the majority of cases the gambler when in the presence of the gaming table, turns towards it as if to ask it for help; for he does not wish to play, or rather having taken his resolution the day before, he thinks of himself still as not wanting to play anymore; he believes in the effectiveness of this resolution. But what he apprehends then in anguish is precisely the total inefficacy of the past resolution. It is there doubtless but fixed, ineffectual, surpassed by the very fact that I am conscious of it. The resolution is still *me* to the extent that I realize constantly my identity with myself across the temporal flux, but it is no longer me—due to the fact that it has become an object for my consciousness. I am not subject to it, it fails in the mission which I have given it. The resolution is there still, I am it in the mode of not-being. What the gambler apprehends at this instant is again the permanent rupture in determinism; it is nothingness which separates him from himself; I should have liked so much not to gamble anymore; yesterday I even had a synthetic apprehension of the situation (threatening ruin, disappointment of my relatives) as *forbidding me* to play. It seemed to me that I had established a *real barrier* between gambling and myself, and now I suddenly perceive that my former understanding of the situation is no more than a memory of an idea, a memory of a feeling. In order for it to come to my aid once more, I must remake it *ex nihilo* and freely. The not-gambling is only one of my possibilities, as the fact of gambling is another of them, neither more nor less. *I must rediscover* the fear of financial ruin or of disappointing my family, etc., I must re-create it as experienced fear. It stands behind me like a boneless phantom. It depends on me alone to lend it flesh. I am alone and naked before temptation as I was the day before. After having patiently built up barriers and walls, after enclosing myself in the magic circle of a resolution, I perceive with anguish that *nothing* prevents me from gambling. The anguish *is me* since by the very fact of taking my position in existence as consciousness of being I make myself *not to be* the past good resolutions which *I am*.

It would be in vain to object that the sole condition of this anguish is ignorance of the underlying psychological determinism. According to such a view my anxiety would come from lack of knowing the real and effective incentives which in the darkness of the unconscious determine my action. In reply we shall point out first that anguish has not appeared to us as a *proof* of human freedom; the latter was given to us as the necessary condition for the question. We wished only to show that there exists a specific consciousness of freedom, and we wished to show that this consciousness is anguish. This means that we wished to established anguish in its essential structure as consciousness of freedom. Now from this point of view the existence of a psychological determinism could not invalidate the results of our description. Either indeed anguish is actually an unrealized ignorance of this determinism— and than anguish apprehends itself in fact as freedom—or else one may claim that anguish is consciousness of being ignorant of the real causes of our acts. In the latter case anguish would come from that of which we have a presentiment, a screen deep within ourselves for monstrous motives which would suddenly release guilty acts. But in this case we should be to ourselves as *things in the world;* we should be to ourselves our own transcendent situation. Then anguish would disappear to give away to *fear,* for fear is a synthetic apprehension of the transcendent as dreadful. . . .

THE SUBJECTIVITY OF VALUES

Now at each instant we are thrust into the world and engaged there. This means that we act before positing our possibilities and that these possibilities which are disclosed as realized or in process of being realized refer to meanings which necessitate special acts in order to be put into question. The alarm which rings in the morning refers to the possibility of my going to work, which is *my* possibility. But to apprehend the summons of the alarm as a summons is to get up. Therefore the very act of getting up is reassuring, for it eludes the question, "Is work *my* possibility?" Consequently it does not put me in a position to apprehend the possibility of quietism, of refusing to work, and finally the possibility of refusing the world and the possibility of death. In short, to the extent that I apprehend the meaning of the ringing, I am already up at its summons; this apprehension guarantees me against the anguished intuition that it is I who confer on the alarm clock its exigency—I and I alone.

In the same way, what we might call everyday morality is exclusive of ethical anguish. There is ethical anguish when I consider myself in my original relation to values. Values in actuality are demands which lay claim to a foundation. But this foundation can in no way be *being,* for every value which would base its ideal nature on its being would thereby cease even to be a value and would realize the heteronomy of my will. Value derives its being from its exigency and not its exigency from its being. It does not deliver itself to a contemplative intuition which would apprehend it as *being* value and thereby would remove from it its right over my freedom. On the contrary, it can be revealed only to an active freedom which makes it exist as value by the sole fact of recognizing it as such. It follows that my freedom is the unique foundation of values and that *nothing,* absolutely nothing, justifies me in adopting this or that particular value, this or that particular scale of values. As a being by whom values exist, I am unjustifiable. My freedom is anguished at being the foundation of values while itself without foundation. It is anguished in addition because values, due to the fact that they are essentially revealed to a freedom, can not disclose themselves without being at the same time "put into question," for the possibility of overturning the scale of values appears complementarily as *my* possibility. It is anguish before values which is the recognition of the ideality of values.

Ordinarily, however, my attitude with respect to values is eminently reassuring. In fact I am engaged in a world of values. The anguished apperception of values as sustained in being by my freedom is a secondary and mediated phenomenon. The immediate is the world with its urgency; and in this world where I engage myself, my acts cause values to spring up like partridges. My indignation has given to me the negative value "baseness," my admiration has given the positive value "grandeur." Above all my obedience to a multitude of tabus, which is real, reveals these tabus to me as existing in fact. The bourgeois who call themselves "respectable citizens" do not become respectable as the result of contemplating moral values. Rather from the moment of their arising in the world they are thrown into a pattern of behavior the meaning of which is respectability. Thus respectability acquires a being; it is not put into question. Values are sown on my path as thousands of little real demands, like the signs which order us to keep off the grass.

Thus in what we shall call the world of the immediate, which delivers itself to our unreflective consciousness, we do not first appear to ourselves, to be thrown subsequently into enterprises. Our being is immediately "in situation"; that is, it arises in enterprises and knows itself first in so far as it is reflected in those enterprises. We discover ourselves then in a world peopled with demands, in the heart of projects "in the course of realization." I write. I am going to smoke. I have an appointment this evening with Pierre. I must not forget to reply to Simon. I do not have the right to conceal the truth any longer from Claude. All these trivial passive expectations of the real, all these commonplace, everyday values, derive their meaning from an original projection of myself which stands as my choice of myself in the world. But to be exact, this projection of myself toward an original possibility, which causes the existence of values, appeals, expectations, and in general a world, appears to me only beyond the world as the meaning and the abstract, logical signification of my enterprises. For the rest, there exist concretely alarm clocks, signboards, tax forms, policemen, so many guard rails against anguish. But as soon as the enterprise is held at a distance from me, as soon as I am referred to myself because I must await myself in the future, then I discover myself suddenly as the one who gives its meaning to the alarm clock, the one who by a signboard forbids himself to walk on a flower bed or on the lawn, the one from whom the boss's order borrows its urgency, the one who decides the interest of the book which he is writing, the one finally who makes the values exist in order to determine his action by their demands. I emerge alone and in anguish confronting the unique and original project which constitutes my being; all the barriers, all the guard rails collapse, nihilated by the consciousness of my freedom. I do not have nor can I have recourse to any value against the fact that it is I who sustain values in being. Nothing can ensure me against myself, cut off from the world and from my essence by this nothingness which I am. I have to realize the meaning of the world and the essence; I make my decision concerning them—without justification and without excuse. . . .

BAD FAITH

The human being is not only the being by whom *négatités* are disclosed in the world; he is also the one who can take negative attitudes with respect to himself. We defined consciousness as "a being such that in its being, its being is in question in so far as this being implies a being other than itself." But now that we have examined the meaning of "the question," we can at present also write the formula thus: "Consciousness is a being, the nature of which is to be conscious of the nothingness of its being." In a prohibition or a veto, for example, the human being denies a future transcendence. But his negation is not explicative. My consciousness is not restricted to *envisioning a négatité*. It constitutes itself in its own flesh as the nihilation of a possibility which another human reality projects as its possibility. For that reason it must arise in the world as a *Not;* it is as a Not that the slave first apprehends the master, or that the prisoner who is trying to escape sees the guard who is watching him. There are even men (for example, caretakers, overseers, gaolers,) whose social reality is uniquely that of the Not, who will live and die, having forever been only a Not upon the earth. Others so as to make the Not a part of their very subjectivity, establish their human personality as a perpetual negation. This is the meaning and function of what Scheler calls "the man of resentment"—in reality, the Not. But there exist more subtle behaviors, the description of which will lead us further into the inwardness of consciousness. Irony is one of these. In irony a man annihilates what he posits within one and the same act; he leads us to believe in order not to be believed; he affirms to deny and denies to affirm; he creates a positive object but it has no being other than its nothingness. Thus attitudes of negation toward the self permit us to raise a new question: What are we to say is the being of man who has the possibility of denying himself? But it is out of the question to discuss the attitude of "self-negation" in its universality. The kinds of behavior which can be ranked under this heading are too diverse; we risk retaining only the abstract form of them. It is best to choose and to examine one determined attitude which is essential to human reality and which is such that consciousness instead of directing its negation outward turns it toward itself. This attitude, it seems to me, is *bad faith (mauvaise foi)*.

Frequently this is identified with falsehood. We say indifferently of a person that he shows signs of bad faith or that he lies to himself. We shall willingly grant that bad faith is a lie to oneself, on condition that we distinguish the lie to oneself from lying in general. Lying is a negative attitude, we will agree to that. But this negation does not bear on consciousness itself; it aims only at the transcendent. The essence of the lie implies in fact that the liar actually is in complete possession of the truth which he is hiding. A man does not lie about what he is ignorant of; he does not lie when he spreads an error of which he himself is the dupe; he does not lie when he is mistaken. . . .

The situation can not be the same for bad faith if this, as we have said, is indeed a lie to oneself. To be sure, the one who practices bad faith is hiding a displeasing truth or presenting as truth a pleasing untruth. Bad faith then has in appearance the structure of falsehood. Only what changes everything is the fact that in bad faith it is from myself that I am hiding the truth. Thus the duality of the deceiver and the deceived does not exist here. Bad faith on the contrary implies in essence the unity of a single consciousness. . . .

Take the example of a woman who has consented to go out with a particular man for the first time. She knows very well the intentions which the man who is speaking to here cherishes regarding her. She knows also that it will be necessary sooner or later for her to make a decision. But she does not want to realize the urgency; she concerns herself only with what is respectful and discreet in the attitude of her companion. She does not apprehend this conduct as an attempt to achieve what we call "the first approach"; that is, she does not want to see possibilities of temporal development which his conduct presents. She restricts this behavior to what is in the present; she does not wish to read in the phrases which he addresses to her anything other than their explicit meaning. If he says to her, "I find you so attractive!" she disarms this phrase of its

sexual background; she attaches to the conversation and to the behavior of the speakcer, the immediate meanings, which she imagines as objective qualities. The man who is speaking to her appears
to her sincere and respectful as the table is round or square, as the
wall coloring is blue or gray. The qualities thus attached to the
person she is listening to are in this way fixed in a permanence like
that of things, which is no other than the projection of the strict
present of the qualities into the temporal flux. This is because she
does not quite know what she wants. She is profoundly aware of
the desire which she inspires, but the desire cruel and naked would
humiliate and horrify her. Yet she would find no charm in a respect
which would be only respect. In order to satisfy her, there must be
a feeling which is addressed wholly to her *personality*—that is, to her
full freedom—and which would be a recognition of her freedom.
But at the same time this feeling must be wholly desire; that is, it
must address itself to her body as object. This time then she refuses
to apprehend the desire for what it is; she does not even give it a
name; she recognizes it only to the extent that it transcends itself
toward admiration, esteem, respect and that it is wholly absorbed
in the more refined forms which it produces, to the extent of no
longer figuring anymore as a sort of warmth and density. But then
suppose he takes her hand. This act of her companion risks changing the situation by calling for an immediate decision. To leave the
hand there is to consent in herself to flirt, to engage herself. To
withdraw it is to break the troubled and unstable harmony which
gives the hour its charm. The aim is to postpone the moment of
decision as long as possible. We know what happens next; the
young woman leaves her hand there, but she *does not notice* that she
is leaving it. She does not notice because it happens by chance that
she is at this moment all intellect. She draws her companion up to
the most lofty regions of sentimental speculation; she speaks of
Life, of her life, she shows herself in her essential aspect—a personality, a consciousness. And during this time the divorce of the body
from the soul is accomplished; the hand rests inert between the
warm hands of her companion—neither consenting nor resisting—a
thing.

We shall say that this woman is in bad faith. But we see immediately that she uses various procedures in order to maintain herself

in this bad faith. She has disarmed the actions of her companion by reducing them to being only what they are; that is, to existing in the mode of the in–itself. But she permits herself to enjoy his desire, to the extent that she will apprehend it as not being what it is, will recognize its transcendence. Finally while sensing profoundly the presence of her own body—to the degree of being disturbed perhaps—she realizes herself as *not being* her own body, and she contemplates it as though from above as a passive object to which events can *happen* but which can neither provoke them nor avoid them because all its possibilities are outside of it. What unity do we find in these various aspects of bad faith? It is a certain art of forming contradictory concepts which unite in themselves both an idea and the negation of that idea. . . .

But consciousness affects itself with bad faith. There must be an original intention and a project of bad faith; this project implies a comprehension of bad faith as such and a pre-reflective apprehension (of) consciousness as affecting itself with bad faith. It follows first that the one to whom the lie is told and the one who lies are one and the same person, which means that I must know in my capacity as deceiver the truth which is hidden from me in my capacity as the one deceived. Better yet I must know the truth very exactly *in order* to conceal it more carefully—and this not at two different moments, which at a pinch would allow us to reestablish a semblance of duality—but in the unitary structure of a single project. How then can the lie subsist if the duality which conditions it is suppressed?

To this difficulty is added another which is derived from the total translucency of consciousness. That which affects itself with bad faith must be conscious (of) its bad faith since the being of consciousness is consciousness of being. It appears then that I must be in good faith, at least to the extent that I am conscious of my bad faith. But then this whole psychic system is annihilated. We must agree in fact that if I deliberately and cynically attempt to lie to myself, I fail completely in this undertaking; the lie falls back and collapses beneath my look; it is ruined *from behind* by the very consciousness of lying to myself which pitilessly constitutes itself well within my project as its very condition. We have here an *evanescent* phenomenon which exists only in and through its own differentiation. To be sure, these phenomena are frequent and we shall see that there is in fact an "evanescence" of bad faith, which, it is evident, vacillates continually between good faith and cynicism: Even though the existence of bad faith is very precarious, and though it belongs to the kind of psychic structures which we might call "metastable," it presents nonetheless an autonomous and durable form. It can even be the normal aspect of life for a very great number of people. A person can *live* in bad faith, which does not mean that he does not have abrupt awakenings to cynicism or to good faith, but which implies a constant and particular style of life. Our embarrassment then appears extreme since we can neither reject nor comprehend bad faith.

THE INADEQUACY OF PSYCHOANALYTIC EXPLANATIONS OF BAD FAITH

To escape from these difficulties people gladly have recourse to the unconscious. In the psychoanalytical interpretation, for example, they use the hypothesis of a censor, conceived as a line of demarcation with customs, passport division, currency control, etc., to re-establish the duality of the deceiver and the deceived. Here instinct or, if you prefer, original drives and complexes of drives constituted by our individual history, make up *reality*. It is neither *true* nor *false* since it does not *exist for itself*. It simply *is,* exactly like this table, which is neither true nor false *in itself* but simply *real*. As for the conscious symbols of the instinct, this interpretation takes them not for appearances but for real psychic facts. Fear, forgetting, dreams exist really in the capacity of concrete facts of consciousness in the same way as the words and the attitudes of the liar are concrete, really existing patterns of behavior. The subject has the same relation to these phenomena as the deceived to the behavior of the deceiver. He establishes them in their reality and must interpret them. There is a *truth* in the activities of the deceiver; if the deceived could reattach them to the situation where the deceiver establishes himself and to his project of the lie, they would become integral parts of truth, by virtue of being lying conduct. Similarly there is a truth in the symbolic acts; it is what the psychoanalyst discovers when he reattaches them to the historical situation of the patient, to the unconscious complexes which they express, to the blocking of the censor. Thus the subject deceives himself about the *meaning* of his conduct, he apprehends it in its concrete existence but not in its *truth,* simply because he cannot derive it from an original situation and from a psychic constitution which remain alien to him.

By the distinction between the "id" and the "ego," Freud has cut the psychic whole into two. I *am* the ego but I *am not* the *id.* I hold no privileged position in relation to my unconscious psyche. I *am* my own psychic phenomena in so far as I establish them in their conscious reality. For example, I am the impulse to steal this or that book from this bookstall. I am an integral part of the impulse; I bring it to light and I determine myself hand-in-hand with it to commit the theft. But I *am* not those psychic facts, in so far as I receive them passively and am obliged to resort to hypotheses about their origin and their true meaning, just as the scholar makes conjectures about the nature and essence of an external phenomenon. This theft, for example, which I interpret as an immediate impulse determined by the rarity, the interest, or the price of the volume which I am going to steal—it is in truth a process derived from self-punishment, which is attached more or less directly to an Oedipus complex. The impulse toward the theft contains a truth which can be reached only by more or less probable hypotheses. The criterion of this truth will be the number of conscious psychic facts which it explains; from a more pragmatic point of view it will be also the success of the psychiatric cure which it allows. Finally the discovery of this truth will necessitate the cooperation of the psychoanalyst, who appears as the *mediator* between my unconscious drives and my conscious life. The Other appears as being able to effect the synthesis between the unconscious thesis and the conscious antithesis. I can know myself only through the mediation of the other, which means that I stand in relation to *my* "id," in the position of the *Other.* If I have a little knowledge of psychoanalysis, I can, under circumstances particularly favorable, try to psychoanalyze myself. But this attempt can succeed only if I distrust every kind of intuition, only if I apply to my case *from the outside,* abstract schemes and rules already learned. As for the results, whether they

are obtained by my efforts alone or with the cooperation of a tech-
nician, they will never have the certainty which intuition confers;
they will possess simply the always increasing probability of scientific hypotheses. The hypothesis of the Oedipus complex, like the
atomic theory, is nothing but an "experimental idea"; as Peirce
said, it is not to be distinguished from the totality of experiences
which it allows to be realized and the results which it enables us to
foresee. Thus psychoanalysis substitutes for the notion of bad faith,
the idea of a lie without a liar; it allows me to understand how it is
possible for me to be lied to without lying to myself since it places
me in the same relation to myself that the Other is in respect to me;
it replaces the duality of the deceiver and the deceived, the essential
condition of the lie, by that of the "id" and the "ego." It introduces
into my subjectivity the deepest intersubjective structure of the *Mit-
sein*. Can this explanation satisfy us?

Considered more closely the psychoanalytic theory is not as
simple as it first appears. It is not accurate to hold that the "id" is
presented as a thing in relation to the hypothesis of the psychoanalyst, for a thing is indifferent to the conjectures which we make
concerning it, while the "id" on the contrary is sensitive to them
when we approach the truth. Freud in fact reports resistance when
at the end of this first period the doctor is approaching the truth.
This resistance is objective behavior apprehended from without: the
patient shows defiance, refuses to speak, gives fantastic accounts of
his dreams, sometimes even removes himself completely from the
psychoanalytic treatment. It is a fair question to ask what part of
himself can thus resist. It can not be the "Ego," envisaged as a
psychic totality of the facts of consciousness; this could not suspect
that the psychiatrist is approaching the end since the ego's relation
to the *meaning* of its own reactions is exactly like that of the psychiatrist himself. At the very most it is possible for the ego to
appreciate objectively the degree of probability in the hypotheses
set forth, as a witness of the psychoanalysis might be able to do,
according to the number of subjective facts which they explain.
Furthermore, this probability would appear to the ego to border on
certainty, which he could not take offence at since most of the time
it is he who by a *conscious* decision is in pursuit of the psychoanalytic therapy. Are we to say that the patient is disturbed by the

daily revelations which the psychoanalyst makes to him and that he seeks to remove himself, at the same time pretending in his own eyes to wish to continue the treatment? In this case it is no longer possible to resort to the unconscious to explain bad faith; it is there in full consciousness, with all its contradictions. But this is not the way that the psychoanalyst means to explain this resistance; for him it is secret and deep, it comes from afar; it has its roots in the very thing which the psychoanalyst is trying to make clear.

Furthermore it is equally impossible to explain the resistance as emanating from the complex which the psychoanalyst wishes to bring to light. The complex as such is rather the collaborator of the psychoanalyst since it aims at expressing itself in clear consciousness, since it plays tricks on the censor and seeks to elude it. The only level on which we can locate the refusal of the subject is that of the censor. It alone can comprehend the questions or the revelations of the psychoanalyst as approaching more or less near to the real drives which it strives to repress—it alone because it alone *knows* what it is repressing.

If we reject the language and the materialistic mythology of psychoanalysis, we perceive that the censor in order to apply its activity with discernment must know what it is repressing. In fact if we abandon all the metaphors representing the repression as the impact of blind forces, we are compelled to admit that the censor must choose and in order to choose must be aware of so doing. How could it happen otherwise that the censor allows lawful sexual impulses to pass through, that it permits needs (hunger, thirst, sleep) to be expressed in clear consciousness? And how are we to explain that it can relax its surveillance, that it can even be deceived by the disguises of the instinct? But it is not sufficient that it discern the condemned drives; it must also apprehend them *as to be repressed,* which implies in it at the very least an awareness of its activity. In a word, how could the censor discern the impulses needing to be repressed without being conscious of discerning them? How can we conceive of a knowledge which is ignorant of itself? To know is to know that one knows, said Alain. Let us say rather: All knowing is consciousness of knowing. Thus the resistance of the patient implies on the level of the censor an awareness of the thing repressed as such, a comprehension of the end toward which the questions of the psychoanalyst are leading, and an act of synthetic connection by which it compares the *truth* of the repressed complex to the psychoanalytic hypothesis which aims at it. These various operations in their turn imply that the censor is conscious (of) itself. But what type of self-consciousness can the censor have? It must be the consciousness (of) being conscious of the drive to be repressed, but precisely *in order not to be conscious of it.* What does this mean if not that the censor is in bad faith?

Psychoanalysis has not gained anything for us since in order to overcome bad faith, it has established between the unconscious and consciousness an autonomous consciousness in bad faith. The effort to establish a veritable duality and even a trinity (*Es, Ich, Ueberich* expressing themselves through the censor) has resulted in a mere verbal terminology. The very essence of the reflexive idea of hiding something from oneself implies the unity of one and the same psychic mechanism and consequently a double activity in the heart of unity, tending on the one hand to maintain and locate the thing to be concealed and on the other hand to repress and disguise it. Each of the two aspects of this activity is complementary to the other; that is, it implies the other in its being. By separating consciousness from the unconscious by means of the censor, psychoanalysis has not succeeded in dissociating the two phases of the act, since the libido is a blind conatus toward conscious expression and since the conscious phenomenon is a passive, faked result. Psychoanalysis has merely localized this double activity of repulsion and attraction on the level of the censor. . . .

[I]n order for me to be able to conceive an intention in bad faith, I must have such a nature that within my being I escape from my being. If I were sad or cowardly in the way in which this inkwell is an inkwell, the possibility of bad faith could not even be conceived. Not only should I be unable to escape from my being; I could not even imagine that I could escape from it. But if bad faith is possible by virtue of a simple project, it is because so far as my being is concerned, there is no difference between being and non-being if I am cut off from my project.

Bad faith is possible only because sincerity is conscious of missing its goal inevitably, due to its very nature. I can try to apprehend myself as *"not being cowardly,"* when I *am so,* only on condition that the "being cowardly" is itself "in question" at the very moment when it exists, on condition that it is itself one question, that at the very moment when I wish to apprehend it, it escapes me on all sides and annihilates itself. The condition under which I can attempt an effort in bad faith is that in one sense, I *am not* this coward which I do not wish to be. But if I were not cowardly in the simple mode of not-being-what-one-is-not, I would be "in good faith" by de-claring that I am not cowardly. Thus this inapprehensible coward is evanescent; in order for me not to be cowardly, I must in some way also be cowardly. That does not mean that I must be "a little" cowardly, in the sense that "a little" signifies "to a certain degree cowardly—and not cowardly to a certain degree." No. I must at once both be and not be totally and in all respects a coward. Thus in this case bad faith requires that I should not be what I am; that is, that there be an imponderable difference separating being from non-being in the mode of being of human reality.

But bad faith is not restricted to denying the qualities which I possess, to not seeing the being which I am. It attempts also to constitute myself as being what I am not. It apprehends me posi-tively as courageous when I am not so. And that is possible, once again, only if I am what I am not; that is, if non-being in me does not have being even as non-being. Of course necessarily I *am not* courageous; otherwise bad faith would not be *bad* faith. But in addition my effort in bad faith must include the ontological com-prehension that even in my usual being what I *am,* I am not it really and that there is no such difference between the being of "being-sad," for example—which I *am* in the mode of not being what I am—and the "non-being" of not-being-courageous which I wish to hide from myself. Moreover it is particularly requisite that the very negation of being should be itself the object of a perpetual nihilation, that the very meaning of "non-being" be perpetually in question in human reality. If I *were not* courageous in the way in which this inkwell is not a table; that is, if I were isolated in my cowardice, propped firmly against it, incapable of putting it in relation to its opposite, if I were not capable of *determining* myself as cowardly—that is, to deny courage to myself and thereby to escape my cowardice in the very moment that I posit it—if it were not on principle *impossible* for me to coincide with my *not-being-courageous* as well as with my being-courageous—then any project of bad faith would be prohibited me. Thus in order for bad faith to be possible, sincerity itself must be in bad faith. The condition of the possibility for bad faith is that human reality, in its most immediate being, in the intrastructure of the pre-reflective *cogito,* must be what it is not and not be what it is. . . .

WE ARE CONDEMNED TO BE FREE

In our attempt to reach to the heart of freedom we may be helped by the few observations which we have made on the subject in the course of this work and which we must summarize here. We established the fact that if negation comes into the world through human-reality, the latter must be a being who can realize a nihilating rupture with the world and with himself; and we established that the permanent possibility of this rupture is the same as freedom. But on the other hand, we stated that this permanent possibility of nihilating what I am in the form of "having-been" implies for man a particular type of existence. We were able then to determine by means of analyses like that of bad faith that human-reality is its own nothingness. For the for-itself, to be is to nihilate the in-itself which it is. Under these conditions freedom can be nothing other than this nihilation. It is through this that the for-itself escapes its being as its essence; it is through this that the for-itself is always something other than what can be *said* of it. For in the final analysis the for-itself is the one which escapes this very denomination, the one which is already beyond the name which is given to it, beyond the property which is recognized in it. To say that the for-itself has to be what it is, to say that it is what it is not while not being what it is, to say that in it existence precedes and conditions essence or inversely according to Hegel, that for it "Wesen ist was gewesen ist"—all this is to say one and the same thing: to be aware that man is free. Indeed by the sole fact that I am conscious of the causes which inspire my action, these causes are already transcendent objects for my consciousness; they are outside. In vain shall I seek to catch hold of them; I escape them by my very existence. I am condemned to exist forever beyond my essence, beyond the causes and motives of my act. I am condemned to be free. This means that no limits to my freedom can be found except freedom itself or, if you prefer, that we are not free to cease being free. To the extent that the for-itself wishes to hide its own nothingness from itself and to incorporate the in-itself as its true mode of being, it is trying also to hide its freedom from itself.

The ultimate meaning of determinism is to establish within us an unbroken continuity of existence in itself. The motive conceived as a psychic fact—that is, as a full and given reality—is, in the deterministic view, articulated without any break with the decision and the act, both of which are equally conceived as psychic givens. The in-itself has got hold of all these "data"; the motive provokes the act as the physical cause its effect; everything is real, everything is full. Thus the refusal of freedom can be conceived only as an attempt to apprehend oneself as being-in-itself; it amounts to the same thing. Human reality may be defined as being such that in its being its freedom is at stake because human reality perpetually tries to refuse to recognize its freedom. Psychologically in each one of us this amounts to trying to take the causes and motives as *things*. We try to confer permanence upon them. We attempt to hide from ourselves that their nature and their weight depend each moment on the meaning which I give to them; we take them for constants. This amounts to considering the meaning which I gave to them just now or yesterday—which is irremediable because it is past—and extrapolating from it a character fixed still in the present. I attempt to persuade myself that the cause is as it was. Thus it would pass whole and untouched from my past consciousness to my present consciousness. It would inhabit my consciousness. This amounts to trying to give an essence to the for-itself. In the same way people will posit ends as transcendences, which is not an error. But instead of seeing that the transcendences there posited are maintained in their being by my own transcendence, people will assume that I encounter them upon my surging up in the world; they come from God, from nature, from "my" nature, from society. These ends ready made and pre-human will therefore define the meaning of my act even before I conceive it, just as causes as pure psychic givens will produce it without my even being aware of them.

Cause, act, and end constitute a *continuum,* a *plenum.* These abortive attempts to stifle freedom under the weight of being (they collapse with the sudden upsurge of anguish before freedom) show sufficiently that freedom in its foundation coincides with the nothingness which is at the heart of man. Human-reality is free because it *is not enough.* It is free because it is perpetually wrenched away from itself and because it has been separated by a nothingness from what it is and from what it will be. It is free, finally, because its present being is itself a nothingness in the form of the "reflection-reflecting." Man is free because he is not himself but presence to himself. The being which is what it is can not be free. Freedom is precisely the nothingness which *is made-to-be* at the heart of man and which forces human-reality *to make itself* instead of *to be.* As we have seen, for human-reality, to be is to *choose oneself;* nothing comes to it either from the outside or from within which it can *receive or accept.* Without any help watsoever, it is entirely abandoned to the intolerable necessity of making itself be—down to the slightest detail. Thus freedom is not a being; it is *the being* of man—that is, his nothingness of being. If we start by conceiving of man as a plenum, it is absurd to try to find in him afterwards moments or psychic regions in which he would be free. As well look for emptiness in a container which one has filled before hand up to the brim! Man can not be sometimes slave and sometimes free; he is wholly and forever free or he is not free at all.

Stuart Hampshire

The style and tone of the English, ordinary-language philosopher Stuart Hampshire are very different from that of Sartre. Yet it can be quite quickly realized that they are discussing the same topics, and indeed coming to somewhat similar conclusions. Both assert the necessary reality of human freedom, and both give conceptual arguments for it, analyzing the nature of intention, action, decision, self-knowledge, prediction, cause, and so forth. Where Sartre seems content to rest with paradox, Hampshire pursues the analysis of what it means to be aware, or fail to be aware, of what one is doing. In arguing the necessary connection between freedom and self-knowledge, he is close to Spinoza and Freud, as well as to Sartre.

These extracts are from Chapter 3 of *Thought and Action* (Chatto & Windus, 1959; in the United States, Humanities Press).

PREDICTION, DECISION, AND FREEDOM OF ACTION

[I]s it logically impossible that a man should predict, rightly or wrongly, what he himself will try to do in the future? Suppose that a fellow prisoner says to me: "I predict that you will try to escape before your time is finished," and that I reply "I am sure that you are right, and I am equally sure that I will fail in the attempt." Upon what principle is the title of "prediction" refused to the first part of the reply, and accorded to the second part? . . . [W]hy should not a man, as a disinterested observer of himself, reflect upon his own record and take his own record as a basis for predicting his own future actions? "I expect that I shall try to escape *at some time*" might be intended more as a prediction than as a statement of intention, if the occasion of the escape is left entirely undetermined. The speaker might be admitting that he has a tendency to try to escape sooner or later, as the evidence of his record shows. A man can examine the record of his own performances disinterestedly, and notice regularities in them, as he would examine the record of another. If the action is located in the undetermined and remote future and does not call for any early decision, a quasi-prediction, or half-prediction, is just possible. The notion of logical possibility, as it is generally used now, is not helpful here: for it only takes one back to the more superficial question of whether we should *call* a proposition of a certain type a prediction or whether we should *call* it the announcement of a decision. We need rather to see the situation that imposes these related distinctions upon us: attempt—achievement and decision—prediction.

There are points in our discourse at which we are compelled to think dialectically, that is, to acknowledge the possibility of an objective contradiction, which arises when two lines of thought, each legitimate within their limits, are pressed too far beyond these limits. The contradiction is objective, in the sense that it does not arise merely from carelessness or ignorance in the use of words. The contradiction here arises from the situation of a speaker speaking about himself as he would commonly speak about others, and simultaneously making a double reference to himself; first, as the observer of himself, who is the author of the statement, and, second, as the independent agent observed. I can make a double reference to myself without contradiction, when I refer to myself, first, as the person observing and, secondly, as the person observed. I may become aware of myself as someone who is trying to annoy somebody else; I suddenly observe myself doing this. But as soon as I become in this way self-conscious about my own *activity,* the situation as I see it, that is, the situation to which my action is adapted, changes. The situation, as viewed by an informed outside observer, has also changed, because of the additional factor of my self-consciousness. Before the moment of self-consciousness, it would not have been true to say that I was deliberately trying to annoy, although this may have been the actual tendency and the likely effect of my actions. But if I still continue on exactly the same course of action after I have become aware of its tendency, it will now for the first time be true to say that I am deliberately trying to annoy. I cannot escape the burden of intention, and therefore of responsibility, which is bestowed upon me by knowledge of what I am doing, that is, by recognition of the situation confronting me and of the difference that my action is making. As soon as I realise what I am doing, I am no longer doing it unintentionally. Any impartial and concurrent awareness of the tendency and effect of my own activities necessarily has to this extent the effect of changing their nature. In virtue of this new awareness, my action may need to be re-described, even though I continue with "the same" activities as before—"the same," that is, when externally viewed, without regard to the intention. That which began as impartial observation turns into something else; the knowledge becomes decision.

When the dialectical movement is viewed from the other side, it appears that, if I suddenly decided to try to annoy someone, after having already acted in a way that was in fact likely to annoy him, the difference represented by my decision would be simply the new knowledge of that which I had previously been doing without this knowledge; the decision here becomes knowledge. I cannot be said to be trying to do anything that I do not know that I am doing. If I can be said to have done something without knowing it, then I must have brought about some result by my activities without having tried to achieve this result. If I know that I am going to do something in the future, I shall do it with a full knowledge of what I am doing, and my present knowledge is then indistinguishable from a decision to do it. Alternatively, my present knowledge is only that, somehow and at some time, a certain result in fact will be the outcome or effect of my activities, without my knowing at the time of acting that this result will be achieved: in which case the achievement will be unintended, and my present knowledge a prediction, based upon some inductive argument. I might have observed that, sooner or later, I always in my life find myself in a certain situation as the effect of my own actions, and on the basis of this evidence I may be certain that this will happen again, even though I do not intend it to happen. I then have a choice, in virtue of this knowledge, of acquiescing in this fatality or of not acquiescing. If I decide actively to prevent this situation arising again from my actions, I will be trying, successfully or unsuccessfully, to avoid bringing about this result. If I am unsuccessful in my serious efforts to prevent it, I will be in a position to claim that I cannot at the present time help bringing about this result. This claim may be disputed by observers, on the ground that I have omitted some step that would in fact have had the desired result. But it would certainly be true that on this occasion I had not allowed the situation to occur of my own free-will, and that in this sense, and to this degree, I was not responsible for the outcome. If I am held responsible, the charge against me is one of incompetence and not of wickedness. If I am successful in my attempt to prevent the situation from arising again, this success has shown predictions based on the past, including possibly my own predictions, to be false. I had made up my mind to try to reverse the trend, without any expectation of succeeding, and yet in fact I succeeded.

It seems therefore that it is through the various degrees of self-consciousness in action, through more and more clear and explicit knowledge of what I am doing, that in the first place I become comparatively free, free in the sense that my achievements either directly correspond to my intentions, or are attributable to my incompetence or powerlessness in execution, which may or may not be venial. Whether and under what conditions the powerlessness is blameworthy is a separate, moral question, or rather set of questions, about which very little can usefully be said in general terms. It plainly depends on the type of incompetence or powerlessness, the type of blame envisaged, on the social purposes that the verdict is designed to serve, and on the particular circumstances of the case.

A man becomes more and more a free and responsible agent the more he at all times knows what he is doing, in every sense of this phrase, and the more he acts with a definite and clearly formed intention. He is in this sense less free the less his actual achievements, that which he directly brings into existence and changes by his activity, correspond to any clearly formed intentions of his own.

It is not by itself a threat to the reality of human freedom that some close observers are able to predict, accurately and with confidence, that which a man is going to do before he actually does it. The threat arises when his own evidently sincere declarations of intention turn out to be comparatively worthless as a basis for predictions of his actions. A sincere declaration of intentions is the most reliable of all sources of information about a man's future action, if he is a free agent, which entails that he is not at the mercy of forces that he does not himself recognise and that are outside his control. This is a necessary truth. If the most reliable basis for prediction of his future actions is the record of similar people in similar situations in the past, and if his own announced decisions afford no basis at all, then he is not free to guide his own activities; he is driven by forces outside his own control. If he in fact generally sets himself to do exactly that which he had intended to do, and if he does not find his activities constantly diverted in a direction that he had not himself designed and thought of, he is fully responsible for his actions, and he is accounted a free agent. He might be a man of inconstant purposes who often changes his mind or hesitates, as new considerations occur to him. He will still be a free, responsible agent as long as his activities at any moment are exactly as he had intended them to be a moment before, even if they are not as he had intended them to be at some earlier time. He is a free agent, in so far as his behaviour is constantly correlated with his evident or declared thoughts and intentions at the time of action rather than with antecedent conditions of some other kind.

The most powerful of all arguments against the reality of human freedom is the old argument that has been given some new basis in some interpretations of the work of Freud and of his followers. It suggests that we commonly do not know what we are trying to do, and are not aware of the tendency of our actions, in a wide domain of our behaviour, and that we can often be brought to recognise the sense of our actions only by the special techniques of psychoanalysis. A neurotic, as described by Freud, is a man who is not fully aware of, and will not acknowledge, the real tendency of his own actions, and who consequently finds that his actual achievements, the constant and forecastable effects of his actions, are generally at variance with his sincerely professed intentions. He is constantly acting in such a way as to defeat his own sincerely professed purposes, until he is brought to acknowledge that he was unconsciously trying to achieve something quite different from his professed aims. In an extended, but not unintelligible, sense of the phrase, he did not know what he was all the time "really" trying to do, or that which he was in a sense trying to do. Because he did not know, he can be said to have been governed by forces outside his own control. The recognition and acknowledgment of them opens for him new possibilities of full self-conscious action; he can now form plans for escaping from his obsessions. He has been brought to recognise internal obstructions to the fulfilment of his sincerely professed intentions, obstructions that he had not recognised before. A psychotic altogether fails to recognise features of the reality around him, and therefore necessarily fails to have any clear knowledge either of what he is trying to do or of the actual effects of his actions. The gap between what he himself honestly says in answer to "What are you trying to do?" and the actual tendency of his actions is so great that he cannot be held in any degree responsible for his achievements. Freud's discovery of the nature of neurosis, which does not involve a total loss of contact with reality, is not easily dismissed from the philosophy of action. We cannot now honestly suppose that the workings of the unconscious mind are to be confined to a clearly marked range of abnormal behaviour. On the contrary, we know that its workings, if they are manifest anywhere, are manifest in a wide range of our behaviour. There is a sense in which we are often not in clear and definite control of ourselves and of our own activities, even when we would ordinarily be said to be completely free agents, if we are judged by the previously recognised criteria. But one must at least be clear in stating the principle behind these recognised criteria before one can decide how far the principle needs to be relaxed and modified in its application. It is possible to view a neurotic's helplessness in the control of his conduct, and his failure to recognise and acknowledge his unconscious aims, as only a great extension of a familiar inability to control conduct, and of failure to act intentionally and with full knowledge of what we are doing. It may be argued that there is no difference of principle: only very great differences of degree upon a scale that is more extended than we had known. The discovery of unconscious purposes may be the discovery that the occasions on which we have, to a greater or less degree, misrepresented to ourselves what we are trying to do are much more common than we had previously believed. Perhaps we were previously misled, in claiming to know what we were trying to do, by ignorance of the fact that, if on these occasions we had been examined further as the psycho-analyst examines, we would have for the first time acknowledged other purposes in ourselves. This would be not a change in the sense of the word "intention," or of the criteria attached to the concept, but rather an empirical discovery about human purposes. The discovery of the unconscious mind, and of its manifestations, may be a discovery of a new range of facts, rather than a proposal for the re-definition of mental concepts. We discover that, applying the accepted criteria of freedom, we are less free, because less fully self-conscious, than we had previously believed. The way to another and deeper level of self-consciousness in behaviour has been opened to us.

At least two conditions are necessary for saying of a man that he is a relatively free agent and responsible for his actions. First, that he generally knows clearly what he is doing; and this condition cannot be fulfilled unless he knows his own situation in the world and generally recognises the relevant features of the situation confronting him at any time. Secondly, there must be a comparatively wide range of achievements open to him, in which he would succeed if he tried, none of which have been made ineligible by human actions and institutions. But are these two conditions jointly sufficient conditions for saying that a man is a free and responsible agent? Or is some further condition required, which in philosophical reflection we are led to believe is never in fact fulfilled, or never could be fulfilled? This question is ordinarily taken to be the problem of free-will. . . .

SELF-KNOWLEDGE AND FREEDOM OF THOUGHT

[I]t may be still argued that the intentions that a man forms, the attempts that he considers making, are limited by his intelligence, his upbringing, the range of his imagination. What a man will *think* of trying to do is narrowly circumscribed and may sometimes be predictable with almost perfect accuracy. The range of his thought, and the possibilities of action that have been suggested to him in his upbringing and social environment, may be exhaustively known. Even if we do not say "He could not have thought of doing that," we certainly say "He could not be expected to think of that." We may justifiably say that only certain projects could possibly occur to him, and, with some intimate knowledge of the man and of his history, we can even now predict precisely what these projects would be. Looking back to an age and a culture remote from our own, we allow that possibilities of action, based on discriminations then unrecognised, were not genuine possibilities of action for those who lived at that time. It seems artificial to speak of a wide range of courses of action as being open to a man, if there is not the slightest possibility (in the sense of likelihood) of his thinking of them as possible courses of action. This objection brings in again the sense of "possibility" that is connected with prediction, and with well-known uniformities in Nature. "It is quite out of the question, and absurd to expect, that the idea of doing that should ever occur to him." Or "it is known that such men as he never do think of doing that." Both these statements are logically compatible with saying "He could (i.e. he would be able to . . .) do that if he chose," and yet both are in many contexts good grounds for saying "It is pointless to blame him for not doing it." There are many things that a man would be able to do if he chose, and that at the same time he cannot be expected to do; it may be evident from general knowledge of human nature and of his social environment that such a thought would never occur to a man of his kind, in such an environment and with such a background. It is merely a fact, and perhaps a fact that could be explained as arising from his history, that the thought and the interest in question would never occur to him. It would be a misleading metaphor to say that he was imprisoned in his own world of thought and within his own circle of interests and that he was not for this reason free; misleading, because a prison is designed to be something from which one will not succeed in escaping even if one tries, to be an insurmountable obstruction. A neurotic may meet an insurmountable obstruction when he tries to dismiss a certain obsessional interest from his mind and finds that he is unable to do it. He may therefore be said to be the prisoner of his obsessional desires, in a metaphor that is only a little overstrained. He may feel himself to be a prisoner, and feel himself unable to escape when he tries. He may feel his will encountering an obstacle whenever he tries; these metaphors seeming to him altogether natural and irreplaceable in describing his experience. But the man whose thoughts and interests in fact revolve within a particular narrow circle is not honestly to be described as imprisoned within this circle, even if we can explain very clearly why his interests are so narrow by reference to his past and to historical causes. In what sense can he be said to be responsible for the limitation of his interests and of his ideas, if we can explain clearly how this limitation arose from causes that were outside his control, such as his upbringing, his education and circumstances of birth? . . .

"He is not the kind of man who would ever think of trying to do that, and he cannot be expected to"—this is a natural exculpating comment upon another man. The degree and kind of the exculpation will depend on the type of explanation suggested of the fact that he never thinks in this way. Does he lack the knowledge or the intelligence and is his critic requiring of him a knowledge and understanding that he does not possess and that he could not have acquired by his own efforts? Or is the situation rather that he does not choose to cultivate certain interests, or to turn his attention to certain possibilities of action, although there is nothing to prevent him doing so? "I am not the kind of man who would ever have thought of trying to do that, and I cannot be expected to"—this is an incomplete form of excuse, as it stands. The question would immediately arise—"Is there anything that prevents you and that makes it impossible for you?" I cannot in good faith say of myself, as I could honestly say of another—"For such and such historical reasons, and under the influence of such and such causes, I have become the kind of man who will always try to do so-and-so rather than so-and-so." Does the phrase "will try" here represent a decision or a prediction? If it is a decision, my character, as a man who will always try to do so-and-so, is something that I have chosen, and the external causes mentioned are only the favouring conditions without which I would not have been in a position to make these decisions. But my character is still something for which I am responsible, chosen from a number of alternatives open to me. If I had thought that my history, and the external causes influencing me, were in any respect bad in their effects, I could at least have tried to find the means to prevent or to minimise their effect. As soon as I have identified these influences as determining influences, I am thereby faced with the choice of acquiescing or of trying to find means of diverting or nullifying their effects. I may excuse my own past, seeing myself now as having been subject to influences that I had not recognised and that I had not brought into consciousness. I judge my own past conduct exactly as I might judge the conduct of another, except that I have a more full, intimate and certain knowledge of the considerations that explicitly guided me than I can normally have of the considerations that guided another. But this is a knowledge of my self-conscious calculations, and not of those explanations of my conduct, which might be given in terms of influences and causes, unrecognised at the time. I cannot have this last kind of knowledge with a policy, a strategy, being required of me as my response to the situation now recognised. If I speak about myself, and the influences bearing on me, in this objective and self-conscious way, whether as an excuse or in simple explanation, I can always be asked when I first recognised these features of my situation. From this moment of recognition onwards, I am responsible for changing or not changing my policies in one direction rather than another. This is one aspect of the asymmetry between proffering excuses for my own conduct and proffering excuses for the conduct of another. I can offer excuses for my own past, and I can explain my own past failures, as if I were speaking of another. I am no longer in the same position as the man who did those things, but am rather the man who understands more fully why they were done. Certainly there is a plain sense in which I am responsible for that which I did at that time, and I can never disclaim responsibility. But I need not think of myself, or judge myself, in any terms and by any standards other than those which I would apply to another. This is indeed the analogy that enables one to think and speak critically of the conduct of others with some assurance, because one is applying terms and standards that one has learnt to apply at first hand in one's own experience. We are always seeing ourselve from the outside, and with the eye of another, in retrospect. But it is unalterably impossible that we should either explain or excuse our present or future voluntary and deliverate action by referring to the influences bearing upon us, unless we are explaining or excusing in advance an expected failure to achieve that which we shall genuinely try to achieve. I can explain why I shall certainly, or almost certainly, fail to achieve an enlargement of my interests and a greater awareness of the range of possibilities open to me; and I can announce my decision to try. But if I try to *predict* that I will not try, and to give the grounds upon which my prediction is based, I shall find myself talking only of the difficulty that I would encounter in bringing myself to make the attempt. It must (logically) always be possible to try, and it must always be possible to achieve *some* extension of the range of my thought, even if this falls short of some ideal that I have conceived. Here the sense of my words necessarily changes when I speak in the first person singular. . . .

[T]here is an artificiality that is almost unavoidable in philosophical discussions of action, because of the tendency to represent human conduct as a set of actions, each of which is a response to a definite situation, as a statement may be an answer to a definite question. A phase of someone's conduct does not consist of a set of actions, each clearly demarcated and unquestionably identifiable, in the sense in which a narrative consists of a set of statements, each clearly demarcated and independently identificable. The situation that confronts a man at any particular time is susceptible of an indefinite set of alternative descriptions. We may therefore think of the situation as the agent would himself identify and describe it as being *the* situation that the agent confronts. But he may of course both misdescribe and misconceive the situation in various different ways. He may suppose that there are features of it that call for action by him, when in fact these features are imaginary and not present at all. The intention that accompanies his action will then incorporate this misapprehension or misdescription. There will then be a wide divergence between the role that he has allotted to himself in his own intentions and his actual performance, even though the performance is entirely voluntary and deliberate. But the relation between thought and action is much more complicated than this, more complicated than has yet been admitted. . . .

"The same action" will sometimes be differently named, and have a different significance, for a man who has beliefs about the existence of things that are different from the agent's beliefs. He will not accept the agent's honest account of his own action as a possible description of any action. Each enclosed within their own systems of belief, they will not find themselves talking about the same action, although they are certainly talking about the same person and the same phase of his activity. Because of this possible doubt about the nature of the action envisaged, moral reflection is often a preliminary, not to a choice between clearly identified alternatives, but to the discovery that a possible course of action, which was originally identified under one description, could be described no less truthfully in quite different terms. For this reason examples of moral dilemmas, occurring in the writings of moral philosophers, must disguise rather than reveal the nature of practical reason. The examples are only identified through their descriptions, and therefore, however exact the description, one is confronted only with abstract possibilities. Each man will recall from his own experience examples of uncertainty in description of his own conduct: there were occasions when his own conduct had not yet assumed the plain outlines which an accurate biography might lend to it. Outside his own experience he can find examples in history: for example, Purcell's Life of Cardinal Manning is a study in the ambiguity of description of conduct, and not a study of hypocrisy. Finally, examples may be found in any fiction that tries to reconstruct exactly this density and uncertainty of experience: in Chekhov and Ibsen and Conrad, who represent their characters as looking forward uncertainly to the possible descriptions of their conduct, to their still unsettled roles.

Aristotle's account of practical reason shows the morally instructed man as *perceiving* the particular actions before him differently from the uninstructed man. The admirable man's superiority does not consist solely in the fact that he is determined, and shows himself in action determined, to be courageous rather than cowardly, noble rather than squalid, but rather in the fact that he immediately recognises in particular cases what the courageous action is. Moral instruction therefore is not so much a strengthening of the will, and an appeal to principle, but rather a habituation that turns a man's interests and intelligence in the right direction, leaving him with a true, but perhaps unstated, conception of what human life should be like. By reference to this habit of exact discrimination he assesses every situation before him. There is no notion in Aristotle's philosophy of "the facts" confronting the agent as by themselves defining the choice before him. A man's own desires, tastes and intentions form the path that he is following and determine the alternatives before him, within the limits set by chance and natural necessity. A man's interests direct his perceptions, and his perceptions pick out the facts that are relevant to his interests.

The freedom of the will is only one part of the freedom of the mind in practical thinking, and injunctions ("You ought to do this") are only one kind of moral judgment. Every rational man must know that his own interests are comparatively confined and that he habitually views situations, and his own performances, largely in the set and conventional terms that he has inherited and passively learnt. He knows that there are many ways of classifying his own actions which would never suggest themselves to him, and that there are many features of the world around him, which he never notices; they would seem to him important, if they did happen to be marked within his habits of classification. He can come to understand the influences in his own environment and upbringing that have formed his habits of thought and his methods of classifying actions, just as he can understand the causes of the fears, aversions and impulses that are obstructions to his will. He will count himself more free in his thought whenever he is able experimentally to detach himself from his own habits and conventions of thought and to describe his own situation and conduct from a new point of view and in new terms. As self-consciousness is a necessary prelude to greater freedom of will, so it is also a necessary prelude to a greater freedom of thought.

Herbert Feigl

Herbert Feigl is one of the many distinguished German-speaking intellectuals who had to flee the Nazi persecution in Europe and subsequently brought great stimulus to America. He is best known for his materialist answer to the mind-body problem—the so-called "identity theory," which says that all mental events are actually physical events in the brain and central nervous system. He presented this at length in his monograph *The Mental and the Physical.*

Feigl seems prepared to accept the results of ordinary-language analysis by philosophers such as Hampshire (and might be persuaded to accept much in Sartre, if demystified), but he insists that a philosophical problem remains about how to relate what we say in everyday terms about intention, action, motives, and so on, to the fast-growing results of neurophysiology. His proposed solution is that these are two different vocabularies which have different meanings and uses, but actually refer to the same set of events. When Feigl's version of the identity theory is compared with Davidson's (in the following reading), it will be noticed that Davidson denies the possibility of the psychophysical laws ("correlations" or "isomorphisms") assumed by Feigl. This may be because Feigl takes sensations or "raw feels" as his paradigm of the mental, whereas Davidson concentrates on beliefs and desires, that is, states with propositional content that serve as reasons for intentional action.

These excerpts are from a paper entitled "Mind-Body, not a Pseudoproblem," in *Dimensions of Mind,* edited by S. Hook (New York University Press, 1960).

THE IDENTITY OF MENTAL
AND NEUROPHYSIOLOGICAL EVENTS

Any serious effort toward a consistent, coherent, and synoptic account of the place of mind in nature is fraught with embarrassing perplexities. Philosophical temperaments notoriously differ in how they react to these perplexities. Some thinkers apparently like to wallow in them and finally declare the mind-body problem unsolvable: *"Ignoramus et ignorabimus."* Perhaps this is an expression of intellectual masochism, or a rationalisation of intellectual impotence. It may of course also be an expression of genuine humility. Others, imbued with greater confidence in the powers of philosophical insight or in the promises of scientific progress, offer dogmatic solutions of the old puzzle. And still others, recognising the speculative and precarious character of metaphysical solutions, and deeply irritated by the many bafflements, try to undercut the whole issue and declare it an imaginary problem. But the perplexities persist and provoke further efforts—often only minor variants of older ones—toward removing this perennial bone of contention from the disputes of philosophers and scientists. Wittgenstein, who tried to "dissolve" the problem, admitted candidly (*Philosophical Investigations,* section 412): "The feeling of an unbridgeable gulf between consciousness and brain process. . . . This idea of a difference in kind is accompanied by slight giddiness," but he added quickly "which occurs when we are performing a piece of logical sleight-of-hand."

As I see it, Wittgenstein's casuistic treatment of the problem is merely one of the more recent in a long line of positivistic (ametaphysical, if not anti-metaphysical) attempts to show that the mind-body problem arises out of conceptual confusions, and that proper attention to the way in which we use mental and physical terms in ordinary language will relieve us of the vexatious problem. Gilbert Ryle, B. F. Skinner, and, anticipating all of them, R. Carnap, have tried to obviate the problem in a similar way. The use of mental or "subjective" terms is acquired by learning the language we all speak in everyday life; this language, serving as a medum of communication among human beings, is by its very nature *intersubjective;* it is on the basis of publicly accessible cues that, for example, the mother tells the child "you feel tired," "now you are glad," "you have a headache," etc., and that the child learns to use such phrases as "feeling tired," "being glad," "having a headache" as applied not only to others, but also to himself when he is in the sort of condition which originally manifested itself in the cues (symptoms, behaviour situations and sequences, test conditions and results, etc.) observable by others. But here is the rub. Even if we *learn* the use of subjective terms in the way indicated, once we have them in our vocabulary we *apply* them to states or conditions to which we, as individual subjects, have a "privileged access." If I report moods, feelings, emotions, sentiments, thoughts, images, dreams, etc., that I experience, I am *not referring to* my *behaviour,* be it actually occurring or likely to occur under specified conditions. I am referring to those states or processes of my direct experience which I live through (enjoy or suffer), to the "raw feels" of my awareness. These "raw feels" are accessible to other persons only indirectly by inference—but it is *myself* who *has* them.

I do not wish to deny that ordinary language serves many purposes quite adequately. As I see it, ordinary language unhesitatingly combines mental (phenomenal) and physical (behavioural) terms in many descriptions and explanations of human and animal conduct or behaviour. "Eagerness was written all over his face"; "He was trembling with anxiety"; "No doubt his gastric ulcer is due to his suppressed hostility"; "An attack of the flu left him in a discouraged and depressed mood for several days"; "A resolute decision finally enabled him to overcome his addiction." As these few illustrations indicate, ordinary language clearly reflects an interactionistic view of the relations of the mental and the physical. As long as we are not too particular about squaring our accounts with the facts established, or at least strongly suggested, by the advances of psycho-physiology, we can manage to keep out of logical troubles. Some philosophers, such as Ryle, Strawson, Hampshire, and other practitioners of the ordinary-language approach, have most persuasively shown that we can talk about the mental life of *"persons,"* i.e. about episodes, dispositions, actions, intentions, motives, purposes, skills, and traits, without getting bogged down in the mind-body puzzles. But, notoriously, there is in this approach scarcely any reference to the facts and regularities of neurophysiology. . . .

The crucial and central puzzle of the mind-body problem, at least since Descartes, has consisted in the challenge to render an adequate account of the relation of the "raw feels," as well as of other mental facts (intentions, thoughts, volitions, desires, etc.) to the corresponding neurophysiological processes. The problems may fairly clearly be divided into scientific and philosophical components. The scientific task is pursued by psycho-physiology, i.e. an exploration of the empirically ascertainable correlations of "raw feels," phenomenal patterns, etc., with the events and processes in the organism, especially in its central nervous system (if not in the cerebral cortex alone). The philosophical task consists in a logical and epistemological clarification of the concepts by means of which we may formulate and/or interpret those correlations.

Scientifically, the most plausible view to date is that of a one-one (or at least a one-many) correspondence of mental states to neurophysiological process-patterns. The investigations of Wolfgang Köhler, E.D. Adrian, W. Penfield, D. O. Hebb, W. S. McCulloch, et al., strongly confirm such a correspondence in the form of an isomorphism of the patterns in the phenomenal fields with the simultaneous pattens of neural processes in various areas of the brain. The philosopher must of course regard this isomorphism as empirically establishable or refutable, and hence as logically contingent. It is conceivable that further empirical evidence may lead the psycho-physiologists to abandon or to modify this view, which on the whole has served so well at least as a fruitful working hypothesis. It is conceivable that some of the as yet more obscure psychosomatic phenomena or possibly the still extremely problematic and controversial "facts" of parapsychology will require emergentist or even interactionistic explanations. (As an empiricist I must at least go through the motions of an "open mind" in these regards!) But tentatively assuming isomorphism of some sort, a hypothesis which is favoured by many "naturalistic" philosophers, are we then to interpret it philosophically along the lines of traditional epiphenomenalism? . . .

Traditionally the most prominent objection to epiphenomenalism has been the argument from the "efficacy of consciousness." We seem to know from our direct experience that moods, pleasure, displeasure, pain, attention, vigilance, intention, deliberation, choice, etc., make a difference in the ensuing behaviour. But, of course, this subjective impression of the causal relevance and efficacy of mental states can easily be explained by the epiphenomenalist: Since, *ex hypothesi,* some dynamically relevant physical conditions are invariably accompanied by mental states, there is, then, also a regular occurence of certain types of behaviour (or of intra-organismic events) consequent upon mental states. For empiricists holding an essentially Humean conception of causality, it is then quite permissible in this sense to speak of the causal efficacy of mental states. There are, it should be noted, countless highly "teleological" processes that occur in our organism evidently without the benefit of any mental influence, guidance, or instigation. For example, the kinds of regenerations and restitutions that are involved in recoveries from many types of physical injury or disease appear as if they were most cleverly "designed," yet for many of these phenomena purely physiological (and perhaps ultimately physico-chemical) explanations are available. Yet according to the epiphenomenalistic doctrine such explanations are sufficient also for behaviour which we ordinarily consider instigated, regulated, or modulated by mental factors. If an effort of concentration facilitates learning algebra, piano-playing, or the like, then consciousness cannot be regarded as a causally irrelevant or superfluous "luxury." I don't think we need to apologise for arguments of this sort. It is true radical Materialists and Behaviourists reject such arguments as "tender-minded," but then radical Materialism or Behaviourism typically *repress* or *evade* the mind-body problem. They do not offer a genuine solution. Epiphenomenalism, while not evading the problem, offers a very queer solution. It accepts two fundamentally different sorts of laws—the usual causal laws and laws of psycho-physiological correspondence. The physical (causal) laws connect the events in the physical world in the manner of a complex network, while the correspondence laws involve relations of physical events with purely mental "danglers." These correspondence laws are peculiar in that they may be said to postulate "effects" (mental states as dependent variables) which by themselves do not function, or at least do not seem to be needed, as "causes" (independent variables) for any observable behaviour.

Laws of concomitance in the physical world could usually be accounted for in terms of underlying *identical* structures. Thus, for example, the correspondence of certain optical, electrical, and magnetic properties of various substances, as expressed in simple functional relations between the refraction index, the dielectric constant, and the magnetic permeability, is explainable on the basis of the atomic structure of those substances. Or, to take a slightly different example, it is in terms of a theory of *one* (unitary) electric current that we explain the thermal, chemical, magnetic, and optical effects which may severally or jointly be used in an "operational definition" of the intensity of the current. Similarily, it is at least a partially successful working programme of psycho-physiology to reduce certain correlated macro-behavioural features to underlying identical neurophysiological structures and processes. It should be emphasised, however, that a further step is needed if we are to overcome the dualism in the epiphenomenalist interpretation of the correlation of subjective mental states with brain states.

The classical attempts in the direction of such unification or of a monistic solution are well known: double-aspect, double-knowledge, twofold-access, or double-language doctrines have been proposed in various forms. The trouble with most of these is that they rely on vague metaphors or analogies and that it is extremely difficult to translate them into straightforward language. I can here only briefly indicate the lines along which I think the "world knot"—to use Schopenhauer's striking designation for the mind-body puzzles—may be disentangled. The indispensable step consists in a critical reflection upon the meanings of the terms "mental" and "physical," and along with this a thorough clarification of such traditional philosophical terms as "private" and "public," "subjective" and "objective," "psychological space(s)" and "physical space," "intentionality," "purposiveness," etc. The solution that appears most plausible to me, and that is entirely consistent with a throughgoing naturalism, is an *identity theory* of the mental and the physical, as follows: Certain neurophysiological terms denote (refer to) the very same events that are also denoted (referred to) by certain phenomenal terms. The identification of the objects of this twofold reference is of course logically contingent, although it constitutes a very fundamental feature of our world as we have come to conceive it in the modern scientific outlook. Utilising Frege's distinction between *Sinn* ("meaning," "sense," "intension") and *Bedeutung* ("referent," "denotatum," "extension"), we may say that neurophysiological terms and the corresponding phenomenal terms, though widely differing in *sense,* and hence in the modes of confirmation of statements containing them, do have identical *referents.* I take these referents to be the immediately experienced qualities, or their configurations in the various phenomenal fields. . . .

I am indeed in agreement with one main line of traditional materialism in that I assume, as does Professor Köhler, that the basic *laws* of the universe are the *physical* ones. But (and this is so brief and crude a formulation that I fear I shall be misunderstood again) this does not commit me in the least as to the nature of the *reality* whose regularities are formulated in the physical laws. This reality is known to us by acquaintance only in the case of our direct experience which, according to my view, is the referent also of certain neurophysiological concepts. And if we are realists in regard to the physical world, we must assume that the concepts of theoretical physics, to the extent that they are instantialised in particulars, are not merely calculational devices for the prediction of observational data, but that they denote realities which are unknown by acquaintance, but which may in some way nevertheless be not entirely discontinuous with the qualities of direct experience. But— "whereof we cannot speak, thereof we must be silent." If this is metaphysics, it seems to me entirely innocuous. I have little sympathy with the mysticism of Eddington or the psychovitalism of Bergson. I reject the former because there is literally nothing that can be responsibly said in a phenomenal language about qualities that do not fall within the scope of acquaintance. Extrapolation will carry us at most to the concepts of unconscious wishes, urges or conflicts as postulated by such "depth-psychologies" as psychoanalysis. And even here, future scientific developments may be expected to couch these concepts much more fruitfully in the language of neurophysiology and endocrinology. And I reject psychovitalism because it involves dualistic interaction. At the very best "intuition" (empathetic imagination) may be heuristically helpful in that it can suggest scientific hypotheses in psychology (possibly even in biology), but these suggestions are extremely precarious, and hence must always be relentlessly scrutinised in the light of objective evidence.

Does the identity theory simplify our conception of the world? I think it does. Instead of conceiving of two realms or two concomitant types of events, we have only one reality which is represented in two different conceptual systems—on the one hand, that of physics and on the other hand, where applicable (in my opinion only to an extremely small part of the world) that of phenomenological psychology. I realise fully that the simplification thus achieved is a matter of *philosophical interpretation*. For a synoptic, coherent account of the relevant facts of perception, introspection, and psychosomatics, and of the logic of theory-construction in the physical sciences, I think that the identity view is preferable to any other proposed solution of the mind–body problem. Call my view metaphysical if you must; I would rather call it *metascientific,* in the sense that it is the result of a comprehensive reflection on the *results* of science as well as on the logic and epistemology of scientific *method*. . . .

Donald Davidson

The work of the American analytical philosopher Donald Davidson on language and mind has attracted intense professional interest in the 1970s. In these parts of his paper "Psychology as Philosophy," in *Philosophy of Psychology,* edited by S. Brown (Macmillan, 1974), he manages to outline a position which offers answers to the three philosophical issues we have been pursuing.

On the mind-body problem, he defends what he calls "anomalous monism," a version of the identity theory which, while accepting that every *individual* mental event must be a physical event (presumably in the brain), denies that there can be any strict psychophysical laws. This denial excludes the identification of *types* of mental events with types of physical events; for instance, the same thought could be embodied in physiologically different brain events.

This irreducibility of the psychological vocabulary (including all the intentional idioms treated by Hampshire and Sartre) to the physical implies, if true, a radical difference between the physical sciences and the social sciences (which we can now *define* as dealing essentially in psychological, intentional terms).

This, in turn, suggests a way out between the horns of the age-old dilemma between freedom and determinism. For even if it is true that every brain event has a sufficient preceding cause and could therefore be predicted in advance (at least in principle), the relevant statement would be expressed in the physical vocabulary. But because of the necessary absence of psychophysical laws, this would not imply any prediction couched in the psychological vocabulary.

Large issues depend, then, on the topic of psychophysical laws. Davidson gives some brief indications here of why he believes there cannot be any—the reasons lying in the *holistic* nature of our intentional concepts, any one of which can be applied to a given person only in the context of applying many others. To give a reason for someone's action, we have to attribute a combination of beliefs and desires to him which would make that action rational.

In a very long historical perspective, Davidson's position can be seen as finding a twentieth-century compromise between dualism and reductive materialism, as Aristotle and Spinoza attempted in their day. But the position is only sketched here; all these issues are under very active discussion in contemporary philosophy.

THE IRREDUCIBILITY OF PSYCHOLOGICAL TO PHYSIOLOGICAL DESCRIPTION, AND OF SOCIAL TO PHYSICAL SCIENCES

Not all human motion is behaviour. Each of us in this room is moving eastward at about 700 miles an hour, carried by the diurnal rotation of the earth, but this is not a fact about our behaviour. When I cross my legs, the raised foot bobs gently with the beat of my heart, but I do not move my foot. Behaviour consists in things we do, whether by intention or not, but where there is behaviour, intention is relevant. In the case of actions, the relevance may be expressed this way: an event is an action if and only if it can be described in a way that makes it intentional. For example, a man may stamp on a hat, believing it is the hat of his rival when it is really his own. Then stamping on his own hat is an act of his, and part of his behaviour, though he did not do it intentionally. As observers we often describe the actions of others in ways that would not occur to them. This does not mean that the concept of intention has been left behind, however, for happenings cease to be actions or behaviour when there is no way of describing them in terms of intention.

These remarks merely graze a large subject, the relation between action and behaviour on the one hand, and intention on the other. I suggest that even though intentional action, at least from the point of view of description, is by no means all the behaviour there is, intention is conceptually central; the rest is understood and defined in terms of intention. If this is true, then what can be said to show that the intentional has traits that segregate it conceptually from other families of concepts (particularly physical concepts) can be applied *mutatis mutandis* to behaviour generally. If the claim is mistaken, then the following considerations apply to psychology only to the extent that psychology employs the concepts of intention, belief, desire, hope, and other attitudes directed (as one says) upon propositions.

Can intentional human behaviour be explained and predicted in the same way other phenomena are? On the one hand, human acts are clearly part of the order of nature, causing and being caused by events outside ourselves. On the other hand, there are good arguments against the view that thought, desire, and voluntary action can be brought under deterministic laws, as physical phenomena can. An adequate theory of behaviour must do justice to both these insights and show how, contrary to appearance, they can be reconciled. By evaluating the arguments against the possibility of determininistic laws of behaviour, we can test the claims of psychology to be a science like others (some others).

When the world impinges on a person, or he moves to modify his environment, the interactions can be recorded and codified in ways that have been refined by the social sciences and common sense. But what emerge are not the strict quantitative laws embedded in sophisticated theory that we confidently expect in physics, but irreducibly statistical correlations that resist, and resist in principle, improvement without limit. What lies behind our inability to discover deterministic psycho-physical laws is this. When we attribute a belief, a desire, a goal, an intention, or a meaning to an agent, we necessarily operate within a system of concepts in part determined by the structure of beliefs and desires of the agent himself. Short of changing the subject, we cannot escape this feature of the psychological; but this feature has no counterpart in the world of physics.

The nomological irreducibility of the psychological means, if I am right, that the social sciences cannot be expected to develop in ways exactly parallel to the physical sciences, nor can we expect ever to be able to explain and predict human behaviour with the kind of precision that is possible in principle for physical phenomena. This does not mean there are any events that are in themselves undetermined or unpredictable; it is only events as described in the vocabulary of thought and action that resist incorporation into a closed deterministic system. These same events, described in appropriate physical terms, may be as amenable to prediction and explanation as any.

320

I shall not argue here for this version of monism, but it may be worth indicating how the parts of the thesis support one another. Take as a first premiss that psychological events such as perceivings, rememberings, the acquisition and loss of knowledge, and intentional actions are directly or indirectly caused by, and the causes of, physical events. The second premiss is that when events are related as cause and effect, then there exists a closed and deterministic system of laws into which these events, when appropriately described, fit. (I ignore as irrelevant the possibility that microphysics may be irreducibly probabilistic.) The third premiss, for which I shall be giving reasons, is that there are no precise psychophysical laws. The three premisses, taken together, imply monism. For psychological events clearly cannot constitute a closed system; much happens that is not psychological, and affects the psychological. But if psychological events are causally related to physical events, there must, by premiss two, be laws that cover them. By premiss three, the laws are not psycho-physical, so they must be purely physical laws. This means that the psychological events are describable, taken one by one, in physical terms, that is, they are physical events. Perhaps it will be agreed that this position deserves to be called *anomalous monism: monism,* because it holds that psychological events are physical events; anomalous, because it insists that events do not fall under strict laws when described in psychological terms.

My general strategy for trying to show that there are no strict psycho-physical laws depends, first, on emphasizing the holistic character of the cognitive field. Any effort at increasing the accuracy and power of a theory of behaviour forces us to bring more and more of the whole system of the agent's beliefs and motives directly into account. But in inferring this system from the evidence, we necessarily impose conditions of coherence, rationality, and consistency. These conditions have no echo in physical theory, which is why we can look for no more than rough correlations between psychological and physical phenomena.

Consider our common-sense scheme for describing and explaining actions. The part of this scheme that I have in mind depends on the fact that we can explain why someone acted as he did by mentioning a desire, value, purpose, goal, or aim the person had, and a belief connecting the desire with the action to be explained. So, for example, we may explain why Achilles returned to the battle by saying he wished to avenge the death of Patroclus. (Given this much, we do not need to mention that he believed that by returning to the battle he could avenge the death of Patroclus.) This style of explanation has many variants. We may adumbrate explanation simply by expanding the description of the action: "He is returning to battle with the intention of avenging the death of Patroclus." Or we may more simply re-describe: "Why is he putting on his armour?" "He is getting ready to avenge Patroclus' death." Even the answer "He just wanted to" falls into the pattern. If given in explanation of why Sam played the piano at midnight, it implies that he wanted to make true a certain proposition, that Sam plays the piano at midnight, and he believed that by acting as he did, he would make it true.

A desire and a belief of the right sort may explain an action, but not necessarily. A man might have good reasons for killing his father, and he might do it, and yet the reasons not be his reasons in doing it (think of Oedipus). So when we offer the fact of the desire and belief in explanation, we imply not only that the agent had the desire and belief, but that they were *efficacious* in producing the action. Here we must say, I think, that causality is involved, i.e. that the desire and belief were causal conditions of the action. Even this is not sufficient, however. For suppose, contrary to legend, that Oedipus, for some dark oedipal reason, was hurrying along the road intent on killing his father, and, finding a surly old man blocking his way, killed him so he could (as he thought) get on with the main job. Then not only did Oedipus want to kill his father, and actually kill him, but his desire caused him to kill his father. Yet we could not say that in killing the old man he intentionally killed his father, nor that his reason in killing the old man was to kill his father.

Can we somehow give conditions that are not only necessary, but also sufficient, for an action to be intentional, using only such concepts as those of belief, desire, and cause? I think not. The reason, very sketchily stated, is this. For a desire and a belief to explain an action in the right way, they must cause it in the right way, perhaps through a chain or process of reasoning that meets standards of rationality. I do not see how the right sort of causal process can be distinguished without, among other things, giving an account of how a decision is reached in the light of conflicting evidence and conflicting desires. I doubt whether it is possible to provide such an account at all, but certainly it cannot be done without using notions like evidence, or good reasons for believing, and these notions outrun those with which we began.

What prevents us from giving necessary and sufficient conditions for acting on a reason, also prevents us from giving serious laws connecting reasons and actions. To see this, suppose we had the sufficient conditions. Then we could say: whenever a man has such-and-such beliefs and desires, and such-and-such further conditions are satisfied, he will act in such-and-such a way. There are no serious laws of this kind. By a serious law, I mean more than a statistical generalization (the statistical laws of physics are serious because they give sharply fixed probabilities, which spring from the nature of the theory); it must be a law that, while it may have provisos limiting its application, allows us to determine in advance whether or not the conditions of application are satisfied. It is an error to compare a truism like "If a man wants to eat an acorn omelette, then he generally will if the opportunity exists and no other desire overrides" with a law that says how fast a body will fall in a vacuum. It is an error, because in the latter case, but not the former, we can tell in advance whether the condition holds, and we know what allowance to make if it doesn't. What is needed in the case of action, if we are to predict on the basis of desires and beliefs, is a quantitative calculus that brings all relevant beliefs and desires into the picture. There is no hope of refining the simple pattern of explanation on the basis of reasons into such a calculus.

Two ideas are built into the concept of acting on a reason (and hence, the concept of behaviour generally): the idea of cause and the idea of rationality. A reason is a rational cause. One way rationality is built in is transparent: the cause must be a belief and a desire in the light of which the action is reasonable. But rationality also enters more subtly, since the way desire and belief work to cause the action must meet further, and unspecified, conditions. The advantage of this mode of explanation is clear: we can explain behaviour without having to know too much about how it was caused.

And the cost is appropriate: we cannot turn this mode of explanation into something more like science.

Explanation by reasons avoids coping with the complexity of causal factors by singling out one, something it is able to do by omitting to provide, within the theory, a clear test of when the antecedent conditions hold. The simplest way of trying to improve matters is to substitute for desires and beliefs more directly observable events that may be assumed to cause them, such as flashing lights, punishments and rewards, deprivations, or spoken commands and instructions. But perhaps it is now obvious to almost everyone that a theory of action inspired by this idea has no chance of explaining complex behaviour unless it succeeds in inferring or constructing the pattern of thoughts and emotions of the agent. . . .

The constitutive force [of rationality] in behaviour derives from the need to view others, nearly enough, as like ourselves. As long as it is behaviour and not something else we want to explain and describe, we must warp the evidence to fit this frame. Physical concepts have different constitutive elements. Standing ready, as we must, to adjust psychological terms to one set of standards and physical terms to another, we know that we cannot insist on a sharp and law-like connection between them. Since psychological phenomena do not constitute a closed system, this amounts to saying they are not, even in theory, amenable to precise prediction or subsumption under deterministic laws. The limit thus placed on the social sciences is set not by nature, but by us when we decide to view men as rational agents with goals and purposes, and as subject to moral evaluation.

Jürgen Habermas

Jürgen Habermas is one of the foremost members of the refounded school of "critical theory" in Frankfurt, West Germany. His distinction between the "empirical-analytic" and the "historical-hermeneutic" sciences would seem to correspond to Davidson's division between the physical and social sciences, for the distinctiveness of the second member of each pair is recognized by both philosophers as lying in the nature of our human understanding of each other's action and speech. But Habermas wants to add a third category, to be called "critical social sciences," which (like Marx and Freud) aim to transform our understanding of ourselves and our society so that change in the world can result. This interest in liberation connects the end of our story with its beginnings in the ancient religious quest.

This brief extract is from his inaugural lecture of 1965, which is published as an appendix to his difficult book *Knowledge and Human Interests* (translated by J. Shapiro, Beacon Press, 1971).

THE IDEA OF CRITICAL SOCIAL SCIENCE

There are three categories of processes of inquiry for which a specific connection between logical-methodological rules and knowledge-constitutive interests can be demonstrated. This demonstration is the task of a critical philosophy of science that escapes the snares of positivism. The approach of the empirical-analytic sciences incorporates a *technical* cognitive interest; that of the historical-hermeneutic sciences incorporates a *practical* one; and the approach of critically oriented sciences incorporates the *emancipatory* cognitive interest. . . .

In the *empirical-analytic sciences* the frame of reference that prejudges the meaning of possible statements establishes rules both for the construction of theories and their critical testing. Theories comprise hypothetico-deductive connections of propositions, which permit the deduction of lawlike hypotheses with empirical content. The latter can be interpreted as statements about the covariance of observable events; given a set of initial conditions, they make predictions possible. Empirical-analytic knowledge is thus possible predictive knowledge. However, the *meaning* of such predictions, that is their technical exploitability, is established only by the rules according to which we apply theories to reality.

In controlled observation, which often takes the form of an experiment, we generate initial conditions and measure the results of operations carried out under these conditions. Empiricism attempts to ground the objectivist illusion in observations expressed in basic statements. These observations are supposed to be reliable in providing immediate evidence without the admixture of subjectivity. In reality basic statements are not simple representations of facts in themselves, but express the success or failure of our operations. We can say that facts and the relations between them are apprehended descriptively. But this way of talking must not conceal that as such the facts relevant to the empirical sciences are first constituted through an a priori organization of our experience in the behavioral system of instrumental action.

Taken together, these two factors, that is the logical structure of admissible systems of propositions and the type of conditions for corroboration, suggest that theories of the empirical sciences disclose reality subject to the constitutive interest in the possible securing and expansion, through information, of feed-back-monitored action. This is the cognitive interest in technical control over objectified process.

The *historical-hermeneutic* sciences gain knowledge in a different methodological framework. Here the meaning of the validity of propositions is not constituted in the frame of reference of technical control. The levels of formalized language and objectified experience have not yet been divorced. For theories are not constructed deductively and experience is not organized with regard to the success of operations. Access to the facts is provided by the understanding of meaning, not observation. The verification of lawlike hypotheses in the empiricial-analytic sciences has its counterpart here in the interpretation of texts. Thus the rules of hermeneutics determine the possible meaning of the validity of statements of the cultural sciences.

Historicism has taken the understanding of meaning, in which mental facts are supposed to be given in direct evidence, and grafted onto it the objectivist illusion of pure theory. It appears as though the interpreter transposes himself into the horizon of the world or language from which a text derives its meaning. But here, too, the facts are first constituted in relation to the standards that establish them. Just as positivist self-understanding does not take into account explicitly the connection between measurement operations and feedback control, so it eliminates from consideration the interpreter's pre-understanding. Hermeneutic knowledge is always mediated through this pre-understanding, which is derived from the interpreter's initial situation. The world of traditional meaning discloses itself to the interpreter only to the extent that his own world becomes clarified at the same time. The subject of understanding establishes communication between both worlds. He comprehends the substantive content of tradition by *applying* tradition to himself and his situation.

If, however, methodological rules unite interpretation and application in this way, then this suggests that hermeneutic inquiry discloses reality subject to a constitutive interest in the preservation and expansion of the intersubjectivity of possible action-orienting mutual understanding. The understanding of meaning is directed in its very structure toward the attainment of possible consensus among actors in the framework of a self-understanding derived from tradition. This we shall call the practical cognitive interest, in contrast to the technical.

The systematic *sciences of social action,* that is economics, sociology, and political science, have the goal, as do the empirical-analytic sciences, of producing nomological knowledge. A critical social science, however, will not remain satisfied with this. It is concerned with going beyond this goal to determine when theoretical statements grasp invariant regularities of social action as such and when they express ideologically frozen relations of dependence that can in principle be transformed. To the extent that this is the case, the *critique of ideology,* as well, moreover, as *psychoanalysis,* take into account that information about lawlike connection sets off a process of reflection in the consciousness of those whom the laws are about. Thus the level of unreflected consciousness, which is one of the intitial conditions of such laws, can be transformed. Of course, to this end a critically mediated knowledge of laws cannot through reflection alone render a law itself inoperative, but it can render it inapplicable.

The methodological framework that determines the meaning of the validity of critical propositions of this category is established by the concept of *self-reflection*. The latter releases the subject from dependence on hypostatized powers. Self-reflection is determined by an emancipatory cognitive interest. Critically oriented sciences share this interest with philosophy.

For Further Reading

On a topic as wide as this one—views of human nature through the ages—a bibliography could go on indefinitely. This will be a short list, mostly of introductory books which will themselves lead one further in various directions.

In most cases, the books from which the readings in this anthology are taken can be recommended to beginners (details have been given in the introductions to each excerpt). The most obvious exceptions are the works of Aristotle, Spinoza, Sartre and Habermas, which are difficult reading even for professional philosophers! For a first approach to Hume, his two *Enquiries* are easier than the *Treatise*. For Marx, the McLellan selection (detailed above) is excellent, but a much shorter collection which is also good is *Karl Marx: Selected Writings in Sociology and Social Philosophy*, edited by T. B. Bottomore and M. Rubel (McGraw-Hill paperback, New York, 1964; Penguin, London, 1963). For Freud, the Pelican Freud Library starting with the *Introductory Lectures on Psycho-Analysis* (Penguin, 1974 onwards) will be more accessible than the *Complete Psychological Works*. In the case of Lorenz, the *Studies in Animal and Human Behavior* are a bit technical; but some of his more popular work is both readable and deep, notably *On Aggression* (Methuen, London, 1966; Bantam Books paperback, New York, 1970). Sartre's *Existentialism and Humanism* (Eyre Methuen, London, 1958; in the United States, Haskell House Publishers) is short and readable, but superficial; before attempting his *Being and Nothingness* read an introduction such as that by Danto in the Fontana Series about to be mentioned.

There are two series that provide short, inexpensive introductions to specific thinkers, most of which can be warmly recommended. In the Fontana Modern Masters series (Viking, New York; Fontana, London) the volumes on Chomsky, Durkheim, Engels, Freud, Jung, Marcuse, Marx, Pavlov, Piaget, Popper, and Sartre are perhaps the most relevant to the present work. In the new Past Masters series (Oxford University Press; in the United States, Hill and Wang), those on Aquinas, Hume, Jesus, and Marx, and those forthcoming on Confucius, Darwin, Engels, Mill, and St. Paul, are the most obviously relevant.

Two good general books are John Passmore's *The Perfectability of Man* (Duckworth, London, 1970; in the United States, Charles Scribner's Sons), which covers a wide range in the history of ideas, and Mary Midgley's *Beast and Man: The Roots of Human Nature* (Harvester Press, Brighton, 1979; Methuen paperback, London, 1980; in the United States, Cornell University Press), which deals with the whole issue of comparison between man and other animals in ethology, psychology, and philosophy.

The exploration of philosophical issues begun in this anthology could continue with Alan Ryan's introductory book *The Philosophy of the Social Sciences* (Macmillan, London, 1970), with the volumes in the Oxford Readings in Philosophy series (Oxford University Press), including *The Philosophy of Social Explanation,* edited by A. Ryan, *The Philosophy of Action,* by A. R. White, and *The Philosophy of Mind,* by J. Glover, and with almost any of the volumes in the Routledge and Kegan Paul series, Studies in Philosophical Psychology.

Permission to reprint is gratefully acknowledged:

St. Thomas Aquinas. Excerpted from the *Summa Theologica*, translated by the Fathers of the English Dominican Province, published by R. & T. Washbourne, Ltd., 1912. Reprinted by permission of the English Province of the Order of Preachers.

Aristotle. Selections from "De Anima" translated by J. A. Smith from *The Oxford Translation of Aristotle* edited by W. D. Ross, vol. 3 (1931). Reprinted by permission of Oxford University Press. Excerpts from "Nicomachean Ethics" translated by W. D. Ross from *The Oxford Translation of Aristotle* edited by W. D. Ross, vol. 9 (1925). Reprinted by permission of Oxford University Press.

The Bible. From the Revised Standard Version of the Bible copyrighted 1946, 1952, © 1971, 1973. Reprinted by permission.

Buddhist Scriptures. From *Buddhist Scriptures*, translated by E. Conze. Reprinted by permission of Penguin Books Ltd.

Chinese Classics. From *Chinese Philosophy in Classical Times*, edited and translated by E. R. Hughes. An Everyman's Library Edition. Reprinted by permission of the publisher in the United States, E. P. Dutton. Published in the United Kingdom by J. M. Dent & Sons Ltd. Publishers. Reprinted by permission.

Noam Chomsky. Excerpted from *Language and Mind*, Enlarged Edition by Noam Chomsky, © 1972 by Harcourt Brace Jovanovich, Inc. Reprinted by permission of the publisher.

Donald Davidson. From "Philosophy as Psychology" by Donald Davidson from *Philosophy of Psychology*, edited by S. C. Brown. Reprinted by permission of Macmillan London and Basingstoke. Published in the United States by Barnes and Noble Books, New York. Reprinted by permission of Barnes and Noble Books, New York.

René Descartes. Excerpted from Descartes, *Philosophical Writings*, edited by Elizabeth Anscombe and Peter Thomas Geach. Reprinted by permission of Thomas Nelson and Sons Limited.

Émile Durkheim. Reprinted from *Selected Writings by Emile Durkheim*, translated by A. Giddens, by permission of Cambridge University Press. Copyright © 1972 by Cambridge University Press.